THE FORTS OF
MAINE

SILENT SENTINELS OF THE PINE TREE STATE

HARRY GRATWICK | FOREWORD BY JOEL EASTMAN

Charleston | London

THE
History
PRESS

Published by The History Press
Charleston, SC 29403
www.historypress.net

Copyright © 2013 by Harry Gratwick
All rights reserved

First published 2013

Manufactured in the United States

ISBN 978.1.60949.536.7

Library of Congress CIP data applied for.

Hang out our banners on the outward walls;
The cry is still, "They come!" Our castle's strength
Will laugh a siege to scorn.

—*William Shakespeare,* Macbeth, *Act V, Scene 5*

CONTENTS

CONTENTS

FOREWORD

Writing a history of the forts of Maine is a daunting undertaking because the state was defended from 1607 through the Cold War. Attempting to include all the defense sites in the state would be an intimidating job whose completion might require a lifetime of work.

Historian Harry Gratwick has taken a much more practical and appealing approach to this task by selecting five different categories of fort sites, which include twelve different fortifications. For each of these forts, he has focused on persons, past and present, connected with these locations.

Once Harry had decided on this pragmatic approach, his research on each fort was meticulous. He located hundreds of published works (books, pamphlets, articles and brochures), visited every site, interviewed scores of persons and found numerous illustrations.

Finally, Harry has taken the immense amount of material he gathered and organized it into interesting and entertaining segments in each of the five chapters. Every person with an interest in Maine's history will find this imaginative work delightful to read and will come away with a plethora of fascinating insights into the history of this remarkable state.

Joel Eastman
Professor Emeritus of History
University of Southern Maine

PREFACE

For this book I have chosen to write about twelve of the several dozen forts that were built in Maine. Some that I have selected are large—others are small. They range from early sixteenth-century forts to the Civil War forts that were built in the middle of the nineteenth century. Geographically, they represent a diverse collection of Maine's forts. In the case of each, I have focused on the actions of a particular individual or individuals whose connections with that fort's history make a particularly interesting story.

ACKNOWLEDGEMENTS

The forts discussed in this book presented several challenges. The first was to select a variety of forts that represented different periods in Maine's history. The next challenge was to choose an interesting person, or persons, connected with each fort's history to write about. The final challenge was to locate each fort and then to find a way to get there, which in some cases was not as easy as it sounds. I would like to thank the following people for helping me with these challenges:

Lee Theriault at Fort Kent; Ken and Patty Higgins, Steve Estes and Glenn Dochermann at Fort McClary; Jonathan Metcalf and Joanne Cameron at Fort Edgecombe; Kelsie Tardiff, Alison Carver and Neill De Paoli at Fort William Henry; Peggy McCrea and Jim Skoglund at Fort St. George; Paige Lilley and Jim Stone at Fort George; Leon Seymour and Dick Dyer at Fort Knox; Gary Morong, Brian Murray and Diane Longley at Fort Popham; Linda Novak and Judy Semple at Old Fort Western; and Joel Eastman, Karen and Harold Cushing and Ken Ford for the Portland Harbor forts (Preble, Scammel and Gorges).

In addition, I appreciate the assistance I received from Gillian Thompson, editor at the *Working Waterfront*; Brenda Steeves (ret.), archivist/special collections, the University of Maine, Orono; Tom Desjardins, historian at the Maine Bureau of Parks and Lands; and Valerie Morton and

Acknowledgements

Linda Whittington at the Vinalhaven Public Library. Finally, I would like to thank my son, Philip, for his historical advice, and as always, I am grateful to my wife, Tita, for her love and support, as well as her editorial expertise.

CHAPTER 1

THE BOUNDARY FORTS

THE COOK AT FORT McCLARY

What was the vice president of the United States doing as a cook at Fort McClary in Kittery, Maine, during the summer of 1864? The short answer is that Hannibal Hamlin's coast guard unit had been called for active duty, and as a member of Company A, he felt obliged to report.

Hamlin had enlisted as a private when the Civil War began. Although Company Commander Llewellyn J. Morse told him that he could have a purely honorary position, Hamlin insisted on active service: "I am the Vice-President of the United States but I am also a private citizen,

Hannibal Hamlin was vice president of the United States when he served as the cook at Fort McClary in the summer of 1864. *Courtesy of Library of Congress, Prints and Photographs Division.*

and as an enlisted member of your company, I am bound to do my duty. I aspire only to be a high private in the rear ranks, and keep step with the boys in blue."

"A Breast Work of Six Guns Shall Be Erected"

Kittery was the first town in Maine to be organized in 1647. In *Maine Forts*, Henry Dunnack writes, "With the coming of the first settlers to the area in 1623, it is possible that individuals built simple breastworks and blockhouses, but the records, if there were any, have disappeared." By the late seventeenth century, Kittery had grown to the point where the town's leading citizen, William Pepperell, built a house for himself on Kittery Point. After he had completed his dwelling, Pepperell purchased an additional twelve acres, which Dunnack tells us was "an ideal place for a fort."

In the early eighteenth century, an intense economic competition had arisen between Maine and New Hampshire, across the Piscataqua River. Customs officials from that colony ordered all ships coming into the harbor to pay a customs duty, which upset Pepperell and other Kittery merchants. Therefore, in 1690, they petitioned the Massachusetts General Assembly to build a fort and station a naval officer in the Port of Kittery.

Thus, Fort William was built in the early eighteenth century. It is not entirely clear whether the fort was named in honor of the elder William Pepperell or the co-ruler of England at the time, King William. Henry Dunnack adds that the purpose of the fortification was to protect merchants from "unreasonable duties imposed by New Hampshire." The directive from the Province of Massachusetts further declared "that six guns with shot and carriages be ordered to the town of Kittery upon their erecting a Breast Work and platform."

Robert Bradley, in the *Forts of Maine 1607–1945*, notes: "What is interesting is that the fort was not built to protect the town from French or Indians, but rather from folks closer at hand. A naval officer was kept in the Port of Kittery to avoid the unreasonable duties exacted from the inhabitants of this province [Maine] passing in and out of the said river [Piscataqua]."

The coming of the American Revolution caused considerable anxiety among the citizens of both Kittery and their neighbor across the river, Portsmouth, New Hampshire. British warships had begun to plunder towns on the Maine coast, and it was feared that the Kittery area would soon be next. Fort William was armed and garrisoned in preparation for an anticipated British attack.

As it turned out, the fortifications so bristled with cannons and bulged with troops that Kittery and Portsmouth were never attacked during the Revolution. After the war, a British naval officer told local officials that he had reconnoitered the town disguised as a fisherman. On his return he informed his superiors that the town was "swarming with soldiers and well defended." The result was that the British plans for an assault were abandoned for the rest of the war.

A Monument to Major McClary

Major Andrew McClary, the leader of a company of New Hampshire militia, was killed at the Battle of Bunker Hill. At the dedication of the Bunker Hill monument, the orator of the day closed with the following tribute to Major McClary:

> *Thus fell Major McClary; the highest-ranking American officer killed at the battle, the handsomest man in the army, and the favorite of the New Hampshire troops. His sun went down at noon on the day that ushered in our Nation's birth.*

The *New Hampshire Gazette*, in its July 1775 issue, printed the following accolade:

> *The Major evinced great intrepidity and presence of mind in the action. His noble soul glowed with ardor and love of his country, and like the Roman Cincinnatus who left his plow, commanded the army, and conquered his opponents, so the Major, upon the first intelligence of hostilities, left his farm and went as a volunteer to assist his suffering brethren where he was soon called to a command which he executed to his eternal honor.*

Fort McClary was officially named in 1808 in honor of Andrew McClary. As relations with Great Britain deteriorated, the federal government completely rebuilt the fort with an upper and lower battery and a barracks and powder magazine. During the War of 1812, it was garrisoned against a possible British threat to the new federal shipyard across the river in Portsmouth.

In fact, Fort McClary and Portsmouth were never attacked because of the difficulty of sailing up the winding Piscataqua River channel. Entering the Piscataqua estuary, an invader would be forced to pass close to Fort McClary before heading up the river to Portsmouth. With many of the fort's cannons located at the water level, an attack would have been suicidal.

Because of worsening relations with Great Britain, the next period of McClary's expansion occurred from 1844 to 1846. This was when the core of the fort, the hexagonal blockhouse (the last to be built in Maine), was built. The blockhouse's unusual construction was, to quote Robert Bradley, "a curious mixture of building materials. The foundation is of mortared field stone; the first story walls are cut granite and the second story is of log construction."

Bradley adds, "What is seen here is the transition in the mid-nineteenth century from earth and timber fortifications to works of stone, a change necessitated by the steady advances in the effectiveness of artillery." In a letter to the *New York Times* in 1857, Oliver Frisbee wrote, "The fort has always been considered a monument to Major McClary. Jefferson Davis visited McClary when he was secretary of war and declared that it was the best-located fort on the Atlantic Coast for defense."

When the Civil War broke out, the Federal government annexed another ten acres of land in the area of the fort, expanding the grounds to twenty-six acres. (No longer did the Kittery Point road run right behind the blockhouse.) The fort was garrisoned, and Congress appropriated a substantial sum to rebuild the blockhouse. A pentagonal granite outer curtain wall was begun, the intention being to completely enclose the grounds of the fort. In addition, a barracks, a cookhouse with a mess hall, a chapel, a hospital, a guardhouse and a magazine—all made of brick—were constructed.

The construction process was extensive. A substantial wharf was built. Derricks were installed to lift granite blocks from incoming ships, and a rail system to move the stones from the harbor area was constructed. There was a granite cutting shop and a finishing shop, as well as a cement shop. Bricks and stone were cut to size. Steve Estes, a member of the Friends of Fort McClary, told me that "it was like putting together a gigantic jigsaw puzzle." Granite was shipped from as far away as Vinalhaven. Other stones were quarried in nearby Marblehead.

The following story, told to me by Steve Estes, has become part of the lore of Fort McClary. At the beginning of the Civil War, the fort was garrisoned by Kittery militia and a few regular army men. (Militia, it should be emphasized, are traditionally less disciplined than regular troops.) One foggy day, when no regulars were on duty, the militia, after their daily ration of rum, sighted what they thought was an enemy ship in the distance and opened fire. The hostile ship, which turned out to be a friendly vessel, fortunately was unharmed. Several weeks later, the militia was removed, and Fort McClary was manned by regular troops for the remainder of the war.

Corporal Hamlin

And so we return to the culinary career of Hannibal Hamlin at Fort McClary. As a distinguished senator from Maine, Hamlin had been used to being an important voice in national politics. As vice president, however, he soon found that his job was largely ceremonial. Indeed, by the summer of 1864, Hamlin was not only "bored to tears," but he was also a lame duck vice president since Abraham Lincoln had chosen Andrew Johnson to be his next running mate. That was fine with Hamlin, who confided to friends his only wish was to be back in the Senate. He could not state this publicly, however, for fear of embarrassing Lincoln and angering his Republican colleagues.

Hamlin and his fellow troopers arrived at Kittery on July 7, and the vice president immediately plunged into the routine of camp life. "We arrived here safely at about nine o'clock and got into quarters about ten. I am very busy," he wrote to his wife, Ellen. "I have been assigned guard duty today, drilled and arranged matters for housekeeping, quarters, etc."

The only concession the vice president asked for was to be housed with the officers. "We are comfortably situated," he wrote Ellen. "The whole

Fort McClary, 1860s. Sketch by C.D. Ordione. *Courtesy of the Maine Historic Preservation Commission.*

company is quartered in the barracks, and it will be to them much like actual service. The duties will not be hard and our sixty days will soon pass off. There will be no difficulty in my getting quarters for you by and by."

After pulling guard duty for three nights, Color Corporal Hamlin (he had been promoted shortly before his arrival) was given the job of company cook. He took the place of an African American named Daniels who had become sick. Hamlin described the situation to his wife:

I do not think I will be able to write much of a letter this morning. Our cook is sick and I am supplying his place. I cooked breakfast and, of course, all hands said it was first rate. Let me see. I made coffee and tea, baked corn fish, fried potatoes and I added a nice Indian loaf from the baker. I have just got my dishes washed and room cleaned and now it is time to begin to get our dinner.

There were no stoves in the kitchen building. Hamlin worked over an open fire in a pit. The vice president, however, loved his job. All the produce was local, and meat, eggs and fish were readily available. The fish were bountiful, as Kittery was a major fishing center. "Having excellent salt water fishing here," he wrote to his son Charles. "Catch Cod, Haddock, Hake, Cunners and Mackerel in abundance."

When he first arrived at Fort McClary in early July, Hamlin's company overlapped for a week with Captain Sylvanus Cobb's company. In *A Memoir of Sylvanus Cobb*, Cobb's daughter, Ella Waite Cobb, wrote that he enjoyed telling the following anecdote about Hamlin: "You who have ever had the good fortune of meeting Hannibal Hamlin under his own vine and fig tree do not need to be told how utterly unused the Vice President was to all outward forms of ceremony."

Ella Cobb continued:

In order to give the VP the fullest possible experience he was detailed for guard duty when Captain Cobb went to and fro from his meals. As the Captain came by on the morning of his departure, Hamlin set down his rifle for a friendly chat instead of giving a salute, which was expected from a soldier to his superior officer. This was too good a chance for Captain Cobb to let pass. With the seeming sternness of a strict disciplinarian he gave Private [Corporal] Hamlin a severe reprimand, which made the boys remember that even though with the removal of one man he would be commander in chief of all the soldiers in the country, as he stood there he was only a corporal and outranked by many of them.

Otherwise, Hamlin's life was anything but that of an ordinary soldier. "I am quartered with the officers in the block house attached to the fort and we get along quite comfortably, but it is becoming somewhat dull," he complained to his wife. "If you Hannie & Frank [his children] could be here, we certainly would have a nice time. My health is excellent. They all say I am growing fat. That ought to be an inducement for you to come."

He continued: "The commodore of the yard promised me the use of his yacht while you might be here and you could not fail to enjoy sails around the harbor and islands on pleasant days with the excitement of catching fish of almost all kinds. I shall, I presume, come home next week, and I am not sure but I shall command you to return with me." (Despite her husband's urgings, there is no evidence that Ellen Hamlin ever visited Fort McClary.)

Hamlin continued to encourage his wife to come: "I can get a girl here to help you take care of the children. Yesterday I had a pleasant ride, to York beach which was very fine, and which I will show you when you come. I do hope you are all over the measles."

After an interlude, Hamlin wrote Ellen on August 17, 1864:

> *I have omitted to write you longer than I intended, but having been quite busy, you must excuse me. The time for our discharge is fast approaching. We will be away in less than three weeks and I shall be glad to go home. I came here only because I deemed it my duty, not because I wanted to come. The soldiers got up a dance last evening in a hall nearby. I should have course have gone, but I went over to hear Mr. Gant of Arkansas make a speech so I did not see or dance with the girls, but I did hear a good speech.*

As his tour of duty drew to a close, politics began to intrude on Hamlin's life, as we see in a letter written to Ellen on August 19, 1864:

> *The State committee has notified me that I must speak in the present election canvass. I hoped sincerely they would let me off, but whether or not they do, I am unwilling to refuse as they would attribute it to my disappointment* [at not being selected as Lincoln's running mate] *which is not the fact. Hence I shall once more make some stump speeches.*

Although his letters to his wife remained upbeat, Hamlin responded more candidly to his son Charles on August 15, 1864, when he revealed his concerns over the failure of Maine's General Howe to receive a promotion:

My Dear son, Just the evening before I started to return to this place [Fort McClary] I received your letter giving me an account of the blunders in the army in your vicinity. Well, I do not know when they are to be remedied; I think I know where the cause lies. I have the highest opinion of General Howe. I esteem him as one of the very best. He should have long since been promoted. I have conferred with the President and Secretary of War until I have become discouraged. It seems to me they do not heed my wishes or recommendations at all. Still I will try and do all I can for the General.

(In fairness to the judgment of President Lincoln and Secretary of War Edwin Stanton, it should be noted that General Albion Parris Howe had a series of contentious relationships with his superior officers that led to his being passed over for division command.)

In early August, Hamlin received the news that there had been a fire in his stables during which some of his horses died. He immediately requested permission to return home. "My loss will be some five or six hundred dollars, but on the whole I am lucky it is no more," he wrote Charles. A week later, Hamlin returned to Kittery to finish his tour of duty, which ended on September 12.

Although he campaigned for the Republican ticket in the fall elections, Hannibal Hamlin spent little time in Washington after his stint at Fort McClary. Following the assassination of Abraham Lincoln in 1865, Andrew Johnson appointed Hamlin collector of customs for the port of Boston. The next year he resigned in protest over Johnson's policies. In 1868, Hamlin was once again elected to his beloved Senate, where he served two more terms until he retired in 1880.

Fort McClary Today

Looking at Fort McClary in the twenty-first century, we see a federal government project that was never completed. Rebuilding was ongoing during the Civil War, but it was discontinued when the war ended. By the early twentieth century, the development of modern artillery made it obsolete. During the Spanish-American War, in 1898, the fort was briefly occupied against a possible attack by a Spanish fleet, and during the First and Second World Wars it was used as an observation post.

When the federal government abandoned the property, nearby residents began to cannibalize the area. They called it "recycling." I was told that

The blockhouse at Fort McClary. The fort's magazine is in the center; the square brick rifleman's house is on the right. *Author's collection.*

many of the granite blocks from the outer walls now serve as mooring blocks for fishing boats in the harbor. To avoid further desecration of the site, the State of Maine purchased Fort McClary from the War Department in 1924 for $31,000.

The history of Fort McClary is interesting because, as with many old fortifications, one is built on the remains of another. To quote Robert Bradley in *The Forts of Maine, 1607–1945*, "The site's history challenges our imagination. Rarely has a locality been fortified against so diverse a series of enemies; from New Hampshire tax collectors, to Britain's Royal Navy, to Confederate raiders to the Spanish Fleet."

THE GODFATHERS OF FORT KENT

There is an indefiniteness to the northeast boundary of the United States as stated in the treaty between this country and Great Britain in 1783.
—Beulah Oxton (1917)

Relations between Great Britain and the United States have not always been as amicable as they have been for the past one hundred years. For much of the nineteenth century, the two countries barely tolerated each other. The American Revolution was followed by Anglo-American hostilities in the War of 1812 and the Aroostook War in 1839–40. Although the British government never recognized the Confederacy during the Civil War, the sympathies of many Englishmen lay with the South.

The Aroostook War

The northeastern boundary of the United States remained ambiguous in the Treaty of Ghent, which ended the War of 1812. The border issue continued to be a problem until 1829, when it was referred to William I of the Netherlands for arbitration. William's decision, however, satisfied neither Great Britain nor the United States, and the boundary controversy simmered. At issue were the rich timberlands located on both sides of the present-day border and the fact that Maine had become a state in 1820 and wanted to clarify its northeastern border.

In the 1830s, lumberjacks from both sides of the disputed territory were harvesting the rich white pine stands in the St. John Valley, known as Madawaska, each claiming they had rights to the area. Maine's governor, Edward Kent, for whom Fort Kent would later be named, had studied the dispute and was convinced that his state's boundary rights were being violated. In the 1838 election, however, John Fairfield defeated Governor Kent. After repeated incidents along the border, Maine's legislature instructed Fairfield to "protect their public lands." Accordingly, they appropriated $800,000 and a draft of ten thousand men.

Meanwhile, the United States Congress allocated $10 million and directed President Martin Van Buren to raise fifty thousand troops to support Maine's position. On the Canadian side of the disputed border, eight hundred fusiliers from Ireland and five hundred British regulars from

Drawing of Fort Kent circa 1842. Fort Kent was built for protection against British Canadians when the boundary line was in dispute. *Collections of Maine Historical Society, Courtesy of Maine Memory Network.*

Quebec prepared to march into Madawaska, the northeastern area between Maine and New Brunswick. War correspondents descended on the border town of Houlton, and the stage was set for conflict.

Insults were hurled back and forth, and a few blows were struck in local taprooms. In the British Parliament, the Duke of Wellington denounced "lawless Yankees of the worst type who were invading and trespassing on British soil." Not to be outdone, Governor Fairfield exhorted his troops: "Fellow Soldiers! An unfounded, unjust and insulting claim of title has been made by the British Government to more than one third of the whole territory of your state." Lieutenant Governor John Harvey of New Brunswick replied by declaring war on the state of Maine on February 13, 1839.

Before and during the Aroostook "War," the Americans and the British Canadians had constructed a series of fortifications in the disputed area. Fort du Petit-Sault was a blockhouse built in 1841 in Edmundson, New Brunswick, which served as an anchor for a series of four military posts that extended to Quebec.

In Maine, Fort Kent and Fort Fairfield were built in 1839. Each occupied a strategic river position. Fort Kent was built at the confluence of the St.

John and Fish Rivers. Booms extended across each river to keep logs from floating downriver to New Brunswick—this triggered a strong reaction from Canadian lumberjacks.

No shots were fired in anger in the almost bloodless Aroostook War, though both sides were ready to fight. Although a few men died of disease, the war's only casualty occurred when a Maine farmer, Nathan Johnston, was killed by "friendly fire" near Fort Fairfield. A troop of soldiers, frustrated by their lack of action, took target practice at the brush behind which Johnson was working in his fields. He died the next day of multiple gunshot wounds.

President Van Buren sent General Winfield Scott to Augusta to mediate the dispute, which was essentially between the State of Maine and the Province of New Brunswick. After an exchange of letters with Governor Harvey, tempers cooled and Maine's troops were withdrawn, although a large civil force remained in the disputed area.

A drawing showing Daniel Webster and Lord Ashburton shaking hands after protracted and difficult negotiations surrounding the determination of the boundary line between the United States and Canada. *Collections of Maine Historical Society, Courtesy of Maine Memory Network.*

Secretary of State Daniel Webster and Alexander Baring (Lord Ashburton) negotiated the Webster-Ashburton Treaty, ratified in 1842. The final settlement seemed to favor the United States. Of the disputed territories, seven-twelfths were given to the United States, whereas five-twelfths were awarded to Great Britain, including lands of marginal economic value. The United States was also granted navigation rights on the St. John River. In retrospect, the British government's goal seems to have been an attempt to improve diplomatic relations with the United States.

The Godfather Was a Major: William Dickey

By the end of the Aroostook War, a logging settlement had sprung up around the Fort Kent blockhouse. It was at this point that the first of Fort Kent's godfathers appeared. William Dickey, the son of a sea captain, was born in 1810 and grew up in the Farmington area of western Maine. As a successful young businessman in the area, he gained a following that resulted in his being elected major of the local militia company. He would be known as "Major" Dickey for the rest of his life.

At the end of the Aroostook War, a special session of the state legislature was convened to consider the Webster-Ashburton Treaty. Dickey was elected to represent the town of Strong, Maine (near Farmington), where he was serving as postmaster. Although the treaty was accepted, Dickey opposed ratification because he felt the "lost territories" belonged to the United States. His views never changed.

Dickey was married and raising a family when he developed a chronic lung disease at the age of forty. As he was convalescing, he remembered a description he had read of the benefits stemming from the bracing climate of Fort Kent. The article was written by a Dr. Greene, who was a regimental surgeon at Fort Kent during the Aroostook War. Dr. Greene claimed that there was no place in the world equal to Fort Kent for curing lung troubles. By this time Dickey was so weak that he could barely walk. At the urging of friends and the state's governor, a former doctor himself, Dickey moved his family to Fort Kent.

On arriving at Fort Kent, the enfeebled Dickey purchased a sawmill and a gristmill and cleared a large tract of farmland. In addition, he built a boardinghouse for his mill and farmhands. Astonishingly, the invigorating climate and cool forest air soon restored his health. William Dickey remained at Fort Kent for five years before moving to Haverhill, Massachusetts, where he opened a manufacturing business. When a fire destroyed his factory, the peripatetic major returned to Fort Kent in 1857. The town remained his home base for the rest of his life.

Dickey volunteered his services to the Union army during the Civil War. He described his activities as "looking after the sanitary condition of Maine boys." At the war's end, Dickey returned to Fort Kent and resumed his life as an entrepreneur and energetic participant in local and state politics.

"Major" Dickey's political life became a whirlwind of activity. During one of his many terms in the Maine House of Representatives, he introduced bills to make new roads. (When he first moved to Fort Kent, there was neither a road, nor bridge nor school within sixty miles of the little village.) Another of Dickey's projects was a proposal to construct bridges at key locations across the St. John River. The Canadian government responded favorably—Congress did not, although eventually four bridges were built.

Although rail service to Aroostook County seemed a remote prospect, Dickey introduced a petition to build a railroad from Houlton to Fort Kent. (The track was completed after his death in 1902.) In all of his proposals, Dickey argued that improved transportation links would enhance the economic development of the area, which, of course, they did.

Over the next three decades, Dickey's constituents came to appreciate his extraordinary vision, energy and achievements. In 1879, the *Aroostook Times* described him as "one of Maine's most prominent Democrats." The following year, as the senior member of the legislature, he was given the honor of calling the House to order on the opening day of the session. Dickey's reelection as a state legislator became an annual event. In an election in 1886, he received every vote in every town in his district.

William Dickey also played a crucial role in determining the future of the Fort Kent blockhouse. Although it was built as a military post, at the end of the Aroostook War a local family purchased and occupied the building for several decades. It is said six children were born there. The town of Fort Kent that grew up around the old blockhouse was incorporated in 1869, doubtless with the assistance of William Dickey. By the 1890s, however, the family had long since departed, and the blockhouse had fallen into disrepair.

Realizing the historic value of the old building, "Major" Dickey introduced a bill "for the preservation of the blockhouse at Fort Kent." The bill passed, and $300 was appropriated for its purchase and renovation. Fort Kent's blockhouse thus became Maine's first state-owned historic fort. A reporter for the *Aroostook Times* wrote: "The bill's sponsor is good for many years. He, if not the blockhouse, is exceedingly well-preserved."

The Fort Kent blockhouse was saved by Major Dickey and is open to the public. *Author's collection.*

Clearly, if not for the actions of "Major" William Dickey, the Fort Kent blockhouse would not be standing today. As such, he deserves to be called the fort's first godfather.

Although it is beyond the scope of the blockhouse story, William Dickey was probably best remembered as the founder of the Madawaska Training School, located in Fort Kent. The school was founded in 1878 to serve as a college for bilingual teachers in the St. John River Valley area. Teachers were trained to educate children of Acadian and French Canadian extraction living in Maine. The success of the school gave Dickey enormous satisfaction, which the school reciprocated. An article in the Bangor paper reported, "The scholars venerate Major Dickey who is a frequent visitor to the school." It should be noted that the Madawaska Training School became the University of Maine at Fort Kent in 1970.

"Major" Dickey served the last of his thirty-three terms in the state legislature in 1897 at the age of eighty-eight. He was as active as ever and sponsored a number of bills and resolutions to benefit the people of his district. The next year, "the venerable patriarch of Fort Kent" began to feel his age. He died at his home on November 19, 1899, eulogized as a man who "contrary to the general course had become broadened and liberal instead of narrow and conservative."

The Godfather Was a Judge: Robert Jalbert

Robert Jalbert was scoutmaster of Fort Kent Troop 189 and a prominent local lawyer and judge. *Courtesy of Lee Theriault.*

After achieving the distinction of being Maine's first historic fort, the Fort Kent blockhouse was promptly forgotten by the state for the next sixty years. Fast-forward to 1959, when thirty-eight-year-old Bob Jalbert, scoutmaster of the reformed Fort Kent Troop 189, moved the scout's meeting place to the old blockhouse, bringing it once again to life.

Like William Dickey, Jalbert was an energetic and extremely capable man. An avid outdoorsman and a registered Maine Guide, Bob had been involved with scouting for thirty-five years. Lee Theriault, the current scoutmaster of Troop 189, said, "Scouting was Bob's passion. If it weren't for Bob, the Fort Kent blockhouse would not be standing today."

Professionally, Jalbert was a successful lawyer who became a respected judge. He was named Citizen of the Year at Fort Kent in 1962. Jalbert was also a past president of the Aroostook County Bar Association and a trustee of his alma mater, the University of Maine at Orono.

Over the years, the once picturesque blockhouse had become a questionable tourist attraction. The building was dirty and filled with trash. When the Boy Scouts moved in, the roof leaked, the walls were unstable and the floor was about to collapse. In sum, it was falling apart. The foresighted

Jalbert realized that fixing the place up would give Troop 189 a home they could be proud of and at the same time provide a significant contribution to the Fort Kent community.

With financial support from the Fort Kent Historical Society, area businessmen and contractors, the scouts completely renovated the fort over a five-year period. The first floor became the epitome of a Boy Scout clubhouse, filled with scouting displays and memorabilia. The second floor was turned into a museum that displayed antique lumbering and farming tools donated by local residents. Outside, the grounds were re-landscaped, numerous trees were planted and an area beside the Fish River was developed for camping.

During the summer, the blockhouse is open daily with scouts conducting tours. On the wall there is a picture of Secretary of the Interior Stuart Udall, who paid a visit in 1961 and was very impressed: "This is marvelous, it is unique, it should become a national shrine. There is nothing else like it." By the summer of 1966, five thousand people were visiting the blockhouse annually. (Today the number runs closer to eight thousand.)

Scout Troop 189 continues to flourish. *Courtesy of Lee Theriault.*

The guest book shows visitors from across the United States, as well as Canada, England, France, Germany, Spain, Mexico, the Caribbean and even India and Australia.

Letters of appreciation have poured in. An excerpt from a Virginia couple who had just finished a canoe trip on the Allagash River is typical:

> *The appearance of The Block House and its Park contrasts greatly with the many historical sites we have seen throughout New England, which have fallen into disrepair. The scout on duty greeted us cordially. He toured the Block House with us and explained its history and was so very gracious that he would reflect great credit on any organization.*

Bob Jalbert stepped aside from the day-to-day operations of Troop 189 in 1965 and became scouting coordinator, though he remained close to the program for the rest of his life. Bob was killed when the small plane he was flying crashed on May 21, 1980. He was returning from a trip to his camp on the Allagash waterway. The plane caught fire, and he died trying to save his friend, Don Michaud, who was flying with him.

Bob's friends today remember him as "a doer." They added, "He was a community minded individual who worked outside the box...He wanted to help the youth of the area...He was strong-minded and feisty...If he wanted something, he got it done...he saved the fort."

A memorial stone honoring Bob and two other deceased leaders of the Fort Kent scouting program, David Daigle and Don Michaud, was unveiled at a ceremony on July 6, 1983. The inscription reads, in part, "to the late Robert Jalbert for his dedicated service to the Fort Kent Blockhouse Park and Troop 189."

Thirty years after his death, Bob Jalbert would be pleased that the scouting program he started at Fort Kent is flourishing. Since Troop 189 was formed in 1959, it has produced over sixty Eagle Scouts, an outstanding achievement. To quote Scoutmaster Lee Theriault, "The scouting tradition in Fort Kent continues to be very popular. Bob's legacy remains with us today."

CHAPTER 2

THE PORTLAND HARBOR FORTS

THE MAJOR AT FORT PREBLE

In April 1861, Major Robert Anderson became a household name as the embattled Union commander of Fort Sumter. Thirteen years previously, Anderson had begun a five-year tour of duty as the commanding officer at Fort Preble in Portland, Maine.

When he arrived at Fort Preble in 1848, Anderson was still recovering from a serious wound he had received in the Battle of Molino del Rey, one of the last engagements in the Mexican War. In the mid-nineteenth century, Fort Preble was mostly used as a troop training center. The War Department saw the post as a place where the convalescing major could recuperate and, at the same time, supervise the training of new recruits.

An "Embargo Fort"

When the British frigate *Leopard* routed the American frigate *Chesapeake* off the Virginia Capes in 1807, war with Great Britain appeared imminent. The First System defenses (primarily earthworks) of the new nation were deemed inadequate. The result was the institution of the Second System of fortifications that were developed by Secretary of War Henry Dearborn during President Thomas Jefferson's second administration.

Fort Preble looking toward Spring Point Ledge Light. Fort Gorges can be seen in the distance on the left. *Author's collection.*

Jefferson's reaction to the British threat was to issue the Embargo Act late in 1807 in an attempt to eliminate trade with Great Britain and thereby avoid potential areas of conflict. By 1808, Congress had passed a number of economic restrictions that New England merchants, including those in Maine, found highly onerous.

One result was the building of Maine's so-called Embargo Forts. In 1808, Secretary of War Dearborn ordered the construction of ten Second System forts (built of brick, stone and earth) to be built along the Maine coast. Included were Forts Preble and Scammel in Portland Harbor. There was a dual purpose to the Embargo Forts, which many saw as an unnecessary federal expense and an infringement on individual freedoms. The stated aim was to guard American harbors and ships from possible British attacks. A less publicized goal was to keep American merchant vessels from leaving their ports, thereby helping to enforce Jefferson's unpopular embargo.

Historian Joshua Smith wrote in an article on Maine's Embargo Forts that the result was "social and political unrest. Riots, kidnappings, gunplay (including one murder), jailbreaks, tarring and feathering and of course rampant smuggling afflicted the District (of Maine)." Smith continued, "To many Mainers there was a direct link between the increased number of fortifications and the Embargo. The first shots fired in anger from several of these posts would not be at foreign invaders but at American citizens engaged in breaking the embargo."

Smith concluded his article by noting: "Far from standing as a symbol of American unity in the face of foreign invasion, these fortifications divided Maine communities and became centers of discord and antagonism. That role would become even more divisive in the approaching War of 1812."

Fort Preble was thus born under a cloud. Joshua Smith tells us that the fort's first commander, Captain Joseph Chandler, worried that the "Enemies of our Country" might try to sneak their ships out of the harbor in defiance of the Embargo. Chandler requested a naval presence in the harbor to discourage this, with the result that USS *Wasp* spent the winter of 1808–1809 in Portland Harbor.

Perhaps it was hoped that by naming the fort in honor of Commodore Edward Preble, local feelings would be calmed. Preble was a Portland native who, as a young man, had distinguished himself in several naval actions during the American Revolution. Twenty-five years later, as a senior officer, Preble's leadership while in command of USS *Constitution* in the Barbary Wars gained international respect for the fledgling United States Navy.

Although the Embargo Act was repealed in 1809, construction on Fort Preble continued until it was completed shortly before the War of 1812. During that war, various units of the regular army and local militia manned the fort. When Winfield Scott and other American soldiers returned from captivity in British Canada, many were in pitiable conditions. While at the fort, a number died in the post's hospital.

Major Anderson at Fort Preble

By the time he arrived at Fort Preble, the forty-three-year-old Anderson had compiled an impressive military record. The 1830s were an active decade for the future commander of Fort Sumter. Highlights included his service as an aide to General Winfield Scott, the United States commander during the Black Hawk War. This was also where Anderson met Captain Abraham Lincoln.

A few years later, Anderson participated in the Florida War against the Seminole Indians, where he was brevetted to captain for his capture of forty-five Indians. It was on this campaign that Anderson contracted the fevers, which were to plague him intermittently the rest of his life.

Late in the decade, Anderson completed a two-year tour of duty at West Point as an instructor of artillery. Ironically, one of his students was P.G.T. Beauregard, who would command the Confederate defenses in Charleston that fired on Fort Sumter in 1861. While at West Point, Anderson translated a French manual on artillery that was published in 1840, with the result that he would eventually be considered an authority on the subject.

When the Mexican War broke out in 1847, Anderson joined the invading army led by his old friend and former commander, Winfield Scott. Declining an offer as an aide on Scott's staff, Anderson fought in a series of battles culminating in the Siege of Vera Cruz. As mentioned, he was badly wounded leading an assault on the enemy lines at Molino del Rey. The citation noted "his gallant and meritorious conduct."

Fort Preble was in the process of being upgraded to a Third System fort by the time Anderson assumed command in 1848. Construction had begun in 1844 and continued intermittently to the end of the Civil War. Additional batteries were added, the fort was enlarged and the remaining earthworks were replaced by masonry fortifications of brick and stone, allowing for multiple-tiered casemented gun emplacements, i.e. more firepower.

The Library of Congress has nineteen boxes of Anderson's papers, including

Major Robert Anderson. *Courtesy of Library of Congress, Prints and Photographs Division.*

four that cover his tour of duty at Fort Preble from 1848

to 1853. We must remember that it was not until the confrontation at Fort Sumter that Anderson became a national hero celebrated in books, songs and purple prose: "Thou hast dashed the cup from treason's hand." The sober soldier who came to Portland in 1848 was a rather reserved career officer who took his job as the fort's commander very seriously. What his correspondence also revealed was, despite his glorious achievements on the battlefield, how much he abhorred war.

Amid the masses of Anderson's military correspondence are the personal letters he wrote to his wife, Eba, whom he had married in 1842. He was clearly a devoted husband and father who missed his family acutely. Apparently, they never came to Maine. "My darling wife and children, you and the youngsters are constantly before me. I am having a lonesome time of it."

While he recovered his health, Anderson served as a teacher and role model for his officers and the enlisted men at the post. When problems came up, Anderson dealt with them directly, as the following examples illustrate:

> On April 27, 1850 Private James Brannan was accused of stealing six dollars from Private Aberl. The case raised several questions including that of military jurisdiction. The two men were drinking at a bar, off the post in Portland, where Private Aberl reportedly got "very drunk." Brannan's defense was that he took Aberl's purse "with the intention of keeping it for him until he should be sober enough to take care of himself." Brannan then apparently got drunk himself and "either spent or lost the six dollars" In the court martial that followed Brannan was found guilty of robbery and imprisoned.

At this point, Anderson stepped in and overruled the court's decision: "I cannot agree with the opinion of the court that the man's (Brannan) appropriating the money to his own use constitutes a felonious act. I consider his failing to return Aberl's money a minor offense, a trespass. The Commanding officer hopes that the Court will on reconsideration see the error of their decision and rectify it," wrote Anderson. Regarding the question of jurisdiction, Anderson added, "Had the act been committed at the Post the accused would have been amenable to military law."

Three years later (February 1, 1853), Anderson again disagreed with the decision of a Fort Preble court-martial that found Private Dennis Curley guilty of desertion. In this case, he felt that the court had not been sufficiently diligent in examining the evidence and had given the man a much lighter sentence. (Whether this was a reaction to his earlier critics is impossible to know.)

The Commanding Officer deeply regrets the necessity, which compels him to differ from the Court on so many material points in this case. He feels the sentence awarded in the case of Curley is inadequate as a punishment for the offence of which he was found guilty. It is merely a withholding of a privilege and is a much lighter sentence than he deserves.

Anderson went on to add, "The proceedings on the second charge are marked with equal if not greater inattention." In this case, the court ignored the testimony of a sergeant who testified that Curley was absent from reveille (morning roll call). In an eight-page letter, Anderson related to his commanding officer at Troy, New York, the conversation he had with two officers (Captain Kilburn and Lieutenant Duncan) from the court who were "very excited" (upset) by his criticism. "Lieutenant Duncan in particular, became quite warm," he wrote. The response Anderson received from regimental headquarters in Troy on March 6 stated that after reviewing the case, his position was validated.

In 1861, there were many who thought that Robert Anderson, a native Kentuckian married to a woman from Georgia, would side with the Confederacy when it came to "managing one of the most delicate assignments in American history." So wrote Adam Goodheart in his book *1861: The Civil War Awakening* when referring to Anderson's command at Fort Sumter. However, the previous examples give us a glimpse into the major's thought process and his commitment to making the right decision militarily, as well as ethically. Goodheart tells us that even Anderson's officers at Fort Sumter did not appreciate "the deeply felt sense of duty and honor…of this pious Kentuckian."

Other Anderson Interests

At the same time that he was managing affairs at Fort Preble, Anderson was actively pursuing two other matters to which he was very committed. The first was the establishment of a military asylum "for the relief and support of invalid soldiers of the army." Anderson explained his idea in a letter to Congress: "If the soldier can look forward, when no longer able to serve his country, to an honorable retreat, to which good conduct and faithful service are the only passport, a retreat where his pride and self respect will not be humbled, enlistments will be made without difficulty."

The bill received strong legislative support, including that of then-senator Jefferson Davis from Mississippi. Davis would become secretary of war in the Pierce administration and, of course, the future president of the Confederacy. He and Anderson had fought together in the Mexican War and were good friends. Davis was properly encouraging when he wrote Anderson, "There is a probability that we will pass a bill to retire officers. There is also a new element created by the grant of extraordinary provisions to invalided soldiers. I am anxious to have your aid knowing that you were so identified with the question."

When the House of Representatives passed the bill in 1851, it was called the Bill of Robert Anderson to Found a House for Old Soldiers. Congratulations poured in to Fort Preble's commander. One legislator wrote, "Dear Major, I congratulate you on the passage of the Army Asylum Bill, believing that you took great interest in it and contributed a good deal by your exertions to forward it."

It took two years to locate a facility, but in 1853 the War Department issued the following statement:

> *The Board have selected the Harrodsburg Springs* [Kentucky] *site for the Asylum and the proceeding will go forward to the President. The place has buildings ready for immediate occupancy and Congress have appropriated $5,000 to cover immediate expenses. Major Robert Anderson is appointed Governor of the Western Military Asylum.*

While at Fort Preble, Anderson also chaired a commission that was charged with updating the United States Army's artillery tactics. (As previously noted, Anderson had become a leading authority on the subject stemming from his translation of the French manual.) With the rather dry title of *The Complete System of Instructions for Siege, Seacoast and Marine Artillery*, the tome was published in 1851. Two thousand copies were printed, and it quickly became the army's official textbook on the subject. Adam Goodheart calls Anderson's manual "a book of such intricately crafted dullness that even a few paragraphs made the unfortunate cadet's head spin and his eyeballs ache." Having browsed through a rough draft of the two-volume set in the Library of Congress, I would agree.

The Civil War and Beyond

With the outbreak of the Civil War, the defenses of Fort Preble and its sister fort, Fort Scammel in Portland Harbor, were again upgraded. Military

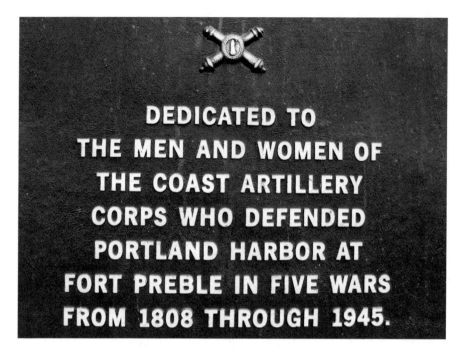

DEDICATED TO
THE MEN AND WOMEN OF
THE COAST ARTILLERY
CORPS WHO DEFENDED
PORTLAND HARBOR AT
FORT PREBLE IN FIVE WARS
FROM 1808 THROUGH 1945.

A plaque at Fort Preble. Can you name the five wars? *Courtesy of Sarah Gratwick.*

historian Joel Eastman tells us that the threat of Confederate steam-powered, ironclad naval vessels and rifled cannons led to the improvement of Portland's harbor defenses. Gun platforms were rebuilt to mount heavier guns, and bombproof ammunition magazines were added. Considerable progress was achieved until the end of the war, when Congress ended funding.

Fort Preble's most noteworthy moment in the war came in July 1863, when Charles Read, a Confederate raider captain, was captured while trying to steal the United States revenue cutter *Caleb Cushing* from Portland Harbor. Read boldly sailed *Caleb Cushing* past Forts Preble and Scammel before he was stopped by two armed steam vessels filled with irate citizens. After a brief exchange of gunfire, the Confederate captain surrendered. Read and twenty-two members of his crew were imprisoned in Fort Preble before being transferred to Boston.

Preble's fortifications continued to be expanded in the 1870s and 1880s. New armament included eight giant Rodman guns in addition to three one-hundred-pound Parrott guns. At the time of the Spanish-American War (1898), Fort Preble also contained two storehouses, a hospital, a guardhouse, an ordnance shed and a headquarters building. Fortunately, the Spanish navy was occupied elsewhere, and the defenses were not tested.

By 1904, Fort Preble had been rebuilt, and four more forts had been constructed in the area to form a combined garrison of 1,600 officers and men. Historian Joel Eastman writes, "The forts featured athletic fields, recreation rooms, libraries, officers' clubs, gymnasiums and tennis courts. The public was invited to watch games against other area forts, as well as civilian teams. The army band and troops marched in the city on national holidays."

In World War I and again in World War II, Fort Preble was manned by reserves, and the fortifications and armaments were upgraded. Donna McKinnon writes in her thesis "Portland Defended: A History of the United States Government Fortifications of Casco Bay, 1790–1945," "The end of World War Two brought also an end to conventional coastal defenses. The Normandy landings proved that large numbers of men could be landed effectively at locations other than ports. Moreover, the atomic bomb and long range high-altitude bombers were weapons beyond the capabilities of harbor defenses."

Fort Preble sat idle until it was decommissioned in 1950. Today, the fort sits on the grounds of Southern Maine Community College, as it has for

Fort Preble from the waterfront today. *Author's collection.*

many years. Although it is not in good repair, portions of Fort Preble's granite outer walls, some concrete structures and the remains of several gun mounts can still be visited. A granite breakwater extends from the fort to the Spring Point Ledge Lighthouse, where one can look out at Forts Scammel and Gorges, guardians—along with Fort Preble—of Portland Harbor in an earlier age.

Postscript

In recognition of his "gallant and meritorious services in defense of Fort Sumter," Robert Anderson was appointed a brigadier general by President Lincoln and put in command of the Department of Kentucky. By 1863, his health from previous wounds and ailments had declined to the point where he was forced to resign from active duty.

After the war, Anderson went to Europe and eventually ended up in southern France, where he hoped the mild climate would restore his well-being. He died in Nice in 1871 at the age of sixty-seven. His body was returned to the United States, where he is buried in the cemetery at West Point.

THE DAUGHTERS OF FORT GORGES

Fort Gorges is a formidable-looking citadel that sits on Hog Island Ledge in the middle of Portland Harbor. The fort was initially part of a War Department plan to strengthen coastal defenses following the War of 1812. Congress, however, did not approve funding for the fort until 1857.

There are several people whose names are closely associated with the history of Fort Gorges. We shall discuss the lives of Annie Murch and Lydia Rust Dahl, the "daughters" who were born at the fort, in due course. But first a look at the seventeenth-century Englishman whose name it honors.

Sir Ferdinando Gorges (1566–1647) was a descendant of Ralph de Gorges, a Norman who came to England with William the Conqueror. Gorges was a loyal Elizabethan soldier who was knighted by his queen for fighting against the Spanish Armada. During his life, he came to know many of the important persons of the age, including Queen Elizabeth and France's King Henry of Navarre.

In the introduction to his biography of Gorges, Professor Richard Preston describes his subject: "Although he was primarily a professional soldier, at various times in his life Gorges was a trader, a privateer, a ship designer, a sailor, a courtier, a country squire and a Justice of the Peace. Above all he was a colonizer and a colonial propagandist."

Gorges has been called the "Father of Maine," even though during the first half of the seventeenth century his various colonial ventures failed, beginning with the

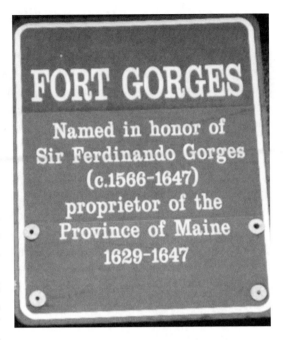

Sign at the entrance to Fort Gorges. *Author's collection.*

ill-fated Popham Colony in 1607. One reason for the lack of success was the modest financial support Gorges was able to provide. Perhaps because of his limited means, Gorges's aims were similar to other colonizers of his era. Like the Elizabethan imperialist Sir Walter Raleigh, Sir Ferdinando instructed his captains to search in their explorations for a New England Eldorado, which would allow him to fund future expeditions.

The climax of Gorges's career came in 1639, when he was granted a royal charter by King Charles I and named proprietor for the Province of Maine. Gorges, however, never got to New England, although he certainly tried. Through no fault of his own, his timing was bad. In the 1620s, war broke out with France and Spain, which upset his plans. And a decade later the now elderly soldier supported the losing (royalist) side in the Puritan Revolution that wracked England. Today, a deserted fort in Portland Harbor is the only place in Maine that bears his name.

Building Fort Gorges

At the end of the War of 1812, it was apparent that most American coastal towns and harbors were still vulnerable to attack. As a result, the War Department began to build a series of large, multistory fortifications of stone or brick called Third System forts. These replaced the First and Second System forts, which were constructed from 1794 to 1812. In most cases, these previous forts were merely glorified earthworks and in constant need of repair.

From 1816 to 1864, more than forty Third System forts were built. Their purpose was to guard important coastal towns, harbors and river entrances against the perceived threat of attack by European powers. It is ironic that the only time Third System forts came into play was during the Civil War. Following the war, most were deemed obsolete due to advances in weaponry.

In Portland, Fort Gorges was built to complement nearby Forts Preble and Scammel by guarding approaches to the inner harbor. When completed and fully armed, the three forts provided layers of overlapping fire for an enemy vessel bold enough to enter the main channel. Fort Gorges was designed to mount fifty-six guns on two lower tiers and an additional thirty-nine guns mounted around the ramparts at the top for a total of ninety-five guns.

Some credit Secretary of War Jefferson Davis with providing the impetus to get the Fort Gorges project approved by Congress. Maine historian Donna McKinnon reminds us that the future president of the Confederacy was no longer secretary of war in the spring of 1857, when he stopped in Portland en route to Bowdoin College to receive an honorary degree.

Congress appropriated $50,000 for a project that began in 1858 and continued through the Civil War. The fort was designed by Colonel Reuben Smart under the supervision of architect Thomas Casey from the Army Corps of Engineers. Fort Gorges was built in the shape of a truncated octagon of six sides with the rear cut off. The final design resembled a smaller version of the brick-built Fort Sumter in Charleston Harbor. It should be noted that both forts faced the sea, which in the case of Fort Sumter made it vulnerable to a land attack.

Since Hog Island Ledge was submerged at high tide, the first order of business was to construct a wharf and build a foundation above the high tide level. Private contractors worked with army engineers to build a wooden barrier around the perimeter of the ledge known as a cofferdam. The seawater was then pumped out, and an area of about an acre and a half was filled with rubble. Essentially, a man-made island had been created on the ledge.

Fort Gorges sits on Hog Island Ledge in the middle of Portland Harbor. *Author's collection.*

The completed foundation provided a workspace, which would become the fort's parade ground at approximately fifteen feet above sea level. This part of the project was completed shortly before the Civil War broke out in April 1861. At this point, an additional appropriation of $100,000 was requested from Congress.

Granite for the walls was shipped from Mount Waldo on the Penobscot River and cut to precise dimensions on nearby House Island, the location of Fort Scammel. Work continued through the war as stonecutters, masons, carpenters and blacksmiths worked together to create a massive, though soon-to-be-obsolete fortress. Some of the workers lived on site, and in 1864, a child was born who was named Annie Gorges Murch.

By 1864, the first two levels of the fort were completed and thirty-four guns were mounted. Fort Gorges was designed to house five hundred men, although the only people who ever lived there were some of the workers and their families. The thought was that in case of a threatened Confederate attack, troops would be sent from nearby Forts Scammel and Preble to man the guns.

The completed Fort Gorges had elements of a medieval fortress with a drawbridge and a sally port at the entrance. (Defenders could "sally forth"

Gun platforms inside the fort. *Author's collection.*

through the sally port to drive off the enemy.) Behind the drawbridge was an enclosed tunnel laced with rifle slits on both sides, leading to the parade ground. (There is some suggestion there may have been a portcullis.) Four sets of apartments were built for officers and their families. It was expected that the enlisted men would "live with their guns."

In addition to officers' quarters there was a bakery, storerooms and latrines. An elaborate piping system carried rainwater to two cisterns in the parade ground and others at the rear (land side) of the fort. The location of magazines for gunpowder was critical. They were built above ground and ventilated by slits in the walls to keep the powder dry. The magazines were placed in the rear of the fort where they were least likely to be hit by enemy shells. The floors were made of wood, with no exposed metal nails that might cause a spark. The magazine walls were whitewashed, which made it easier to see what was inside. Incidentally, the outside walls were also whitewashed so attackers would realize the harbor was defended.

By the end of the Civil War, the original plans for Fort Gorges were largely completed. During the war, however, changes in military technology had made Third System forts like Fort Gorges obsolete. The fifteen-inch

Rodman gun, which could fire a 435-pound projectile four miles, had been developed and was superior to land batteries previously in use. Rifled guns were mounted on steam-powered armored vessels with destructive hitting power. Work had begun to modernize Fort Gorges's defenses by adding a sod-covered roof to protect the lower gun encasements and powder magazines when Congress cut off further funding in 1877.

Two Daughters and Two Grandfathers

Two baby girls were born at Fort Gorges, separated by an interval of forty-one years. It appears, not surprisingly, that as adults neither of the women was aware of the other's existence. The first to be born was Annie Gorges Murch, granddaughter of Reuben Smart, the architect who designed Fort Gorges. Smart had a well-deserved reputation as an expert builder, having constructed forts in several southern states, as well as in Maine. An acquaintance once said, "Anything Reuben Smart does is built on a foundation of granite and honor."

Smart was working on Fort Knox on the Penobscot River when he was ordered to Portland to supervise the building of Fort Gorges. The story goes that Smart arrived at Hog Island Ledge as the first load of granite was being delivered. There he met his old friend and colleague Captain Thomas Casey of the Army Corps of Engineers, who had overall responsibility for the project.

The two men had previously collaborated to build a number of New England lighthouses and forts, which included Forts Scammel and Preble in Portland Harbor and Forts Popham and Knox farther up the coast. After the war, the brilliant Casey, who had graduated first in his class at West Point, would go on to an outstanding career with the corps. In Washington, D.C., alone he supervised the construction of the Library of Congress and the State, War and Navy Building (now the Eisenhower Executive Office Building). He also completed construction of the long-delayed Washington Monument.

Reuben Smart must have been pleased that his son-in-law, John Murch, was an accomplished mason and stonecutter. Murch had married Reuben Smart's daughter, Valeria, and the two men enjoyed working together on the Fort Gorges job. Five years into the project, Annie Gorges Murch was born on March 18, 1863, in a little cottage on the Fort Gorges building site.

Annie Murch was three months old when her father made a name for himself. The occasion was the closest Portland would come to a Confederate attack. On the night of June 26, 1863, Murch spotted a ship, which turned out to be the captured fishing vessel *Archer*, slipping into Portland Harbor in a daring attempt to capture the Federal revenue cutter *Caleb Cushing*. Murch observed that there appeared to be "trouble aboard" and rowed to nearby Fort Preble to sound the alarm. The raid, led by Confederate captain Charles Read, is a story in itself. Murch, however, deserves credit for alerting the authorities who would capture Read before he could complete his mission.

Unfortunately, John Murch died of typhoid fever the next year. When her father died, Annie went to live with her mother's family in South Portland. The little girl grew up never forgetting her connection with Fort Gorges. In a newspaper interview she gave shortly before her death in 1939, Annie Murch reported growing up listening to her grandmother talk about what a brave man her father was. She came to revere her family name. "I vowed never to change it," she said. "And I never did."

True to her word, Annie Murch married a distant cousin, George Murch, from Hamden, Maine. George was interested in streetcars, and shortly after their marriage the couple moved to Salem, Massachusetts, where he began a career as a trolley car engineer and manager. George worked on numerous projects throughout the eastern United States and became an expert in installing electric trolley lines. In all, George Murch built and managed thirty-seven different lines before his death in 1929.

Annie and George had two sons, one of whom died in an unspecified accident in 1900. Following her husband's death, Annie Murch lived with her other son, Wilbur George, and his two children on Peaks Island, across from Fort Gorges. Wilbur George Murch carried on the family tradition and became an expert mason. Jobs were hard to come by in the 1930s during the Depression, but George was able to work on a WPA project at Fort Preble, repairing the casements his grandfather, Reuben Smart, had built years before.

Annie Murch, the "first" Fort Gorges baby, died at her home on Peaks Island in 1939 at the age of seventy-six. Although she admitted she "hadn't been back to the fort in years," Annie was always very proud of the fact that her father and grandfather were an important part of the history of Fort Gorges.

The "second" Fort Gorges baby, Lydia Rust, was born on July 8, 1904. At the end of the Spanish-American War, her grandfather, Charles Rust, was hired as caretaker of the deteriorating fort. Newspaper reports of her birth described Lydia as "the fort's only native." By this time Annie Murch was married and no longer living in Maine, so it is understandable why the two women never knew about each other.

Coincidentally, Lydia was also very young when her father died. She was only two, though she and her mother spent the next ten years living at Fort Gorges until her grandfather died in 1916. At that point, she moved to South Portland. As an older adult, Lydia, now married to John Dahl, described the fort:

> *Most people probably can't imagine living out there now since it has been torn up by vandals. But my family had beautiful quarters. Our apartment was a wonderful place to live. We lived in one of the old officers' quarters with heavy hardwood floors, papered walls, huge fireplaces and copper-framed casement windows. There were also extensive recreational facilities*

Lydia Dahl with Portland's Neighborhood Youth Corps director, Nelson Pepin. *Courtesy of Portland Herald.*

including a bowling alley. We grew fruit trees and you could pick roses from our dining room window. Tourists and friends were coming and going all the time in those days.

Although Lydia had happy memories of life at the fort, she was also a lonely child who looked forward to school on the mainland so she could play with other children. "We'd run up a red flag over the fort's wall as a signal that I was ready to go to school," she recalled. She also remembered Sunday dinners with "dozens of guests." On the menu were lobsters from the family's traps, some caught by Lydia.

Unlike Annie Murch, Lydia Rust Dahl returned to Fort Gorges as an older woman. She served as an informal consultant in the 1960s for Portland's Neighborhood Youth Corps, which was using the fort as a training center while reconditioning its interior. "I'll do anything I can to help," Dahl said. "I love to talk about the old fort. It seems every time we get together with friends we always end up talking about those days." Nelson Pepin, director of the youth program, said, "It was a great help having Mrs. Dahl tell us what all those rooms were used for."

Lydia was married twice, first to Elmer Sanborn, who died in 1950, and then to John Dahl, who died in 1988. She had a total of six children, two of whom predeceased her. Lydia died in February 1996 at the age of ninety-one.

"And Don't Be Startled by the Pigeons"

During the Spanish-American War, Fort Gorges was used as a storage depot for mines, or "torpedoes" as they were called. A steel-framed building was erected on the parade ground to keep mines and mine anchors. The mines were laid at the various entrances to Portland Harbor to prevent an attack by enemy warships.

The Army Corps of Engineers continued to use the fort as a storehouse for mines and other explosives until the end of World War II. After the war, the fort was neglected and pillaged by vandals. Fort Gorges was acquired by the City of Portland in 1960 and placed on the National Register of Historic Places, although the city did little to maintain it.

Fort Gorges has been the scene of sporadic activities in the last fifty years. As previously noted, the Neighborhood Youth Corps that consulted with Lydia Dahl in the 1960s did some cleaning up and restoration. Developers in the 1980s were briefly interested in restoring the fort and

Outline of torpedo storage shed on parade ground used in the Spanish-American War. *Author's collection.*

turning it into a tourist attraction with a museum and a restaurant. And over the years, the Casco Bay Island Development Association (CBIDA) has held dances at the site.

Today, Fort Gorges is a town park and the southernmost point of the Maine Island Trail. It is open to the public though accessible only by boat. Large boats can use the wharf at high tide. Kayakers and smaller craft can land at any tide. Visitors are welcome but are reminded that they explore the walls and inner chambers of the fort at their own risk. And don't be startled by the pigeons that are nesting everywhere.

THE SAVIOR OF FORT SCAMMEL

The granite stones from Fort Scammel would probably be holding back the waters of Casco Bay today if it were not for the persistence of a woman named Hilda Cushing. In 1955, the owner was preparing to

49

View of Portland Harbor from Fort Preble, 1853. House Island can be seen just left of center; the building is Fort Scammel. *Courtesy of Maine Memory Network.*

sell the stones from the decaying fort for the construction of the Spring Point Ledge breakwater when Hilda appeared. But we are getting ahead of ourselves.

Fort Scammel has an interesting history. The original fort was built in 1808 on the western end of House Island in Casco Bay. It consisted of a semicircular brickwork wall and was sited to command the main channel into Portland Harbor. At the rear of the fort was a wooden blockhouse that served as a barracks. The brickwork of the fort and the blockhouse were whitewashed to make it clear to the enemy that the harbor was defended

A Revolutionary War Hero

Alexander Scammel grew up in Massachusetts and graduated from Harvard College in 1769. When the Revolution broke out, he was reading law for John Sullivan in Boston. The Revolution forced Sullivan to close his law practice and form a regiment, which Scammel joined as a major. Early in the Revolution, Scammel fought with George Washington in the Battles of Trenton and Princeton and reportedly rallied his retreating troops at a critical point in the latter battle. After being promoted to colonel, Scammel's reputation as a field commander was further enhanced by his actions at the important Battle of Saratoga in 1777.

After Saratoga, he was appointed adjutant general to the Continental army, whose resolve was sorely tested by the harsh winter at Valley Forge. Scammel became a member of Washington's inner circle for the remainder of the war and was known for his easy manner and ability to lighten the often-grim mood of the general's staff. It is said that Washington considered him one of the funniest men in the army.

In 1780, Washington ordered Scammel to oversee the trial and subsequent execution of British major John Andre, an assignment that he thoroughly disliked. Andre had been captured carrying incriminating papers from Benedict Arnold that revealed the defenses of West Point to the British. Andre was convicted of espionage and was hanged on October 2, 1780.

Alexander Scammel was commanding a New Hampshire light infantry regiment at the siege of Yorktown in 1781 when he was shot from ambush near Williamsburg, Virginia. He died from his wounds twelve days later on October 6. George Washington called him "a man who inspired us to do our duty."

A monument was erected to Scammel at Williamsburg, and the Alexander Scammel Bridge, which crosses the Bellamy River near Durham, New Hampshire, was dedicated in 1933. Twenty-seven years after his death, the gallant officer was honored for his military service to the nation with a fort in Casco Bay, which has stood since 1808.

The Evolution of Fort Scammel

General Scammel's career was far more active than that of his namesake on House Island in the middle of Portland Harbor. During the War of 1812, Fort Scammel served as one of several fortifications defending Casco Bay. There is some question as to whether shots were ever exchanged with the British. Historian Joshua Smith, however, has located an article in a contemporary Boston newspaper stating that early in August 1813 Fort Scammel fired on a privateer lurking outside the harbor. Both sides had been posturing, and the Americans may have been saying: "We dare you to come in." In any case, after an exchange of gunfire, no further advances were made.

In the middle of the nineteenth century, Fort Scammel was modernized as part of a seacoast defense program known as Third System fortifications.

The walls were extended to allow for the mounting of more and larger cannons. Eventually, the walls completely enclosed the fort, making it less vulnerable to a naval attack or a landing by enemy troops. The blockhouse was removed in 1860 in anticipation of further expansion.

When the Civil War broke out, engineers drew up plans for massive changes. Work continued intermittently, and the fort was in a nearly completed state when Congress cut off further appropriations in 1876. Exact cost figures are difficult to estimate, but the total expense seems to have been well under $1 million.

Famed army engineer Thomas Lincoln Casey supervised Fort Scammel's "upgrade." In today's parlance, we would call the general a brilliant multitasker. No matter how you phrase it, Casey was overextended. It has already been noted that when he began rebuilding Fort Scammel he was also working on Fort Gorges and Fort Preble in Portland Harbor. At the same time, he was also in charge of "improving" Fort Knox on the narrows of the Penobscot River at Bucksport and Fort Popham at the mouth of the Kennebec River.

At the age of thirty, Casey had already achieved an impressive reputation as an engineer. Fort Scammel was probably the least known of his assignments, but it proved to be one of his most challenging. By the time he began work on the fort in 1862, Casey had developed a skilled cadre of workers. He taught them how to anchor foundations in the tidal waters of the submerged ledges of coastal Maine and how to build derricks and unload heavy equipment in nearly inaccessible places.

Casey's workers at Fort Scammel would need to utilize all of their skills. Immense granite blocks the size of an automobile had to be landed and put into place. To anchor stones on ledges in the face of ten-foot tides required not only skill but also, in the words of a contemporary journalist, "a touch of building magic."

Most of Fort Scammel is located underground, where narrow granite stairways and damp passageways lead to the various levels. The subterranean architecture, however, is far from ponderous and dismal. There are graceful archways, and one of the spiral stairways that leads down to a drinking well has been described as "a true architectural beauty."

The whole effect of a walk through the cavernous chambers of Fort Scammel reminds one of an underground cathedral. There is a particularly majestic quality to the Great Magazine. Although the fort was never garrisoned, General Casey constructed it to accommodate three hundred men. A *Maine Sunday Telegram* article describing the vast

The west bastion was one of three added by army engineer Thomas Casey during the Civil War era. *Courtesy of Karen Lannon.*

munitions magazine as "a hall where a hundred couples might dance" is not an exaggeration.

Perhaps the closest Fort Scammel ever came to actual warfare was during the Spanish-American War, when it was manned by six fifteen-inch Rodman guns and used as the center for laying harbor mines. Later, during the First World War, emplacements for antiaircraft guns were installed, although in keeping with its past history, Fort Scammel was never attacked.

Waiting for Hilda

Fort Scammel was saved from demolition in the 1950s when Hilda Cushing purchased House Island. Before this, however, the island went through several metamorphoses. Hilda's son, Captain Hal Cushing, told me that until 1905 there was a prosperous fishing village and an extensive fish-drying operation at the end of the island opposite the fort. A pretty little graveyard, the former post cemetery, sits in the middle of the figure-eight-

The "Savior of Fort Scammel," Hilda Cushing.
Courtesy of Karen Lannon.

shaped island, with the graves of twenty-two residents who lived on House Island in the nineteenth century. In addition, three members of the Cushing family are buried in the cemetery: Hilda, her husband Harold and a daughter, Heidi.

Several years after the Spanish-American War, the fishing village was torn down, and an elaborate United States Immigration and Quarantine Station was built. During this period, from 1907 to 1937, House Island was known as the Ellis Island of the North. The quarantine station was busiest in the early 1920s following the adoption of the Emergency Quota Act, which restricted the number of immigrants who could enter the country. In November 1923, the ships *President Polk* and *George Washington* were diverted from New York City to Portland, and 218 immigrants from those ships were quarantined at the station.

Immigration officials considered House Island an ideal location. A Grand Trunk Railway station was located at the Portland docks, allowing easy access by rail to North American cities for immigrants arriving in Portland. The Grand Trunk system operated primarily in the eastern Canadian provinces and New England states.

William Husband, commissioner general of United States Immigration, described why he preferred House Island to other locations: "It was secure and the whole island was available instead of those detained being obliged to go out under guard with only a few patches of green grass upon which they might set foot, as at some other places."

The brick detention barracks that once housed six hundred people has been demolished, but the remainder of the original 1907 buildings are standing, including the doctor's house, the detention barracks and the quarantine hospital, which today is the Cushing family house on the island.

During Prohibition in the 1920s, House Island was used as a transfer area for rumrunners. When his mother bought the island, Captain Hal remembers finding scores of broken bottles scattered along the beach. One of the more apocryphal stories that Hal remembers hearing as a child was the one about a hermit who lived in one of the abandoned shacks on a pier at the eastern end of the island. The story was that he had two dogs—he liked the one named Rosie the best because she washed the dishes.

There are also stories that tell of House Island, specifically the now-deserted Fort Scammel, being used for fraternity initiations in the 1930s. Pledges were brought from the former Gorham State Teacher's College, which is today a part of the University of Maine system. Apparently, the extensive underground chambers of the fort were used to frighten the uninitiated, who were made to walk the plank in one of the dungeons.

When the War Assets Administration was selling off unused bases after World War II, the Fort Scammel end of House Island was purchased by retired navy captain Lincoln King. Interestingly, Captain King paid $1,200 for the fort, the same price the government bought the land for in 1807. A newspaper report quotes him as saying, "It is simply marvelous. It couldn't be replaced for a million dollars." A few years later, King would sell his "marvelous" fort to Hilda Cushing for a good deal less than $1 million.

House Island Tours

Hilda Cushing was a Mainer from Portland. She was married to Harold Cushing and living with her family in South Portland in the 1950s when she heard that Captain King was planning to sell Fort Scammel. King had learned that the U.S. Army Corps of Engineers was going to build a breakwater from the shore to the Spring Point Ledge Lighthouse. Since he was moving to California, King decided to sell stones from his fort for the project, which would require fifty tons of granite. (The lighthouse was built in 1897, and the Army Corps of Engineers were planning to construct a nine-hundred-foot-long granite breakwater.)

Hilda was appalled by the idea that the fort would be torn down, and after getting in touch with King, she began to negotiate a purchase price with him. Hilda believed strongly in the importance of the past and always told her children, "You can't have tomorrow without having yesterday."

According to her son, Captain Hal, Hilda's discussions with King went on for over a year. He remembers as a child going with his mother and sister, Karen, to King's house in Paris, Maine, and hearing his mother negotiate with King. In 1954, Hilda learned that the retired captain was leaving for the West Coast. She met him at the Portland train station with a check and told him, "This is what I can pay. If you accept my offer, cash it when you get to California."

Hilda never told her children what her offer was, though it must have been satisfactory because Captain King got off the train in Boston and cashed the check. Years later, King's granddaughter came east and told the Cushings, "It was a great offer," though to this day Captain Hal and his sister, Karen Lannon, do not know what the exact price was.

It should be emphasized at this point that Hilda had only bought half of House Island. A man named Stanley Pettengill owned the rest, which included the remaining buildings from the former Immigration Station. When Pettengill heard that Hilda was looking for an island house, he offered to sell her his half of the island. Pettengill was getting married, and the idea of living on an island in the middle of Casco Bay did not appeal to his fiancée. She gave him an ultimatum: "Make up your mind—the island or me." Pettengill chose his fiancée, and that is how Hilda Cushing became the owner of the eastern half of House Island in 1955 and the first person to own the twenty-four-acre island in its entirety.

Once Hilda had purchased the island, she told her family, "We have to figure out how to pay the taxes." This would become an ongoing and all-consuming family experience. First, they had to clean up the entire island. Captain Hal told me that when they purchased the island, the vandalism was so bad that all the windows (except one pane of glass) were broken. The family spent the next few years rehabilitating the buildings and preparing to make the island pay for itself.

Before she bought Fort Scammel, Hilda had rebuilt and sold houses on the mainland to returning servicemen in the 1940s. In fact, during the course of her life she was well known throughout the Portland area for her business savvy. During the course of her life, Hilda bought and sold real estate, boats and houses. She also raised, showed and sold horses. Hilda Cushing was thus well equipped to turn House Island into a profit-making venture.

If Hilda is the "Savior of Fort Scammel," then her children, Captain Hal Cushing and Captain Karen Lannon, should be called the keepers of

The "Keepers of Fort Scammel," Captains Karen Lannon and Hal Cushing. *Author's collection.*

the fort. To help pay island expenses, Hilda, Hal and Karen began hosting lobster dinners and offering tours of Fort Scammel in 1960.

House Island Tours has become a full-service operation. Tours of Fort Scammel and lobster dinners continue to be major attractions. In addition, they hosted family reunions, 4-H groups, wedding parties, ship commissionings, school reunions, bus tours, company outings, birthday parties and various evening events. Karen Lannon told me that on a given week in the summer they would host fifty events.

People from all over the United States have taken tours of House Island. As I walked around the twenty-four-acre island, I was impressed with the rolling fields, mowed paths and the large tent under which one hundred guests could be seated at picnic tables. Visitors to House Island depart from Long Wharf off of Commercial Street in Portland and are transported in the passenger vessel *Chippewa*, operated by Captain Hal and Karen.

During the summer of 2008, the bicentennial of Fort Scammel was celebrated with an event called "Behind the Wall," which offered tours of the fort twice a week. It had been fifty-two years since Hilda had saved the fort from demolition and turned House Island into a successful business operation. In the process, family members had researched Fort Scammel's history at the National Archives and led thousands of people on tours of the fortress.

For the bicentennial celebration, the plan was to manicure the earthworks and offer tours of the fort. Historian Joel Eastman, Ken Thompson and several members of the Coast Defense Study Group (CDSG) and Portland Harbor Museum assisted in the project by helping to clear the grounds and acting as tour guides.

In the process of removing years of accumulated earth, vegetation and undergrowth, a clearer profile of the fort was revealed. A narrow-gauge railroad leading under the fort to an experimental cement mixer was found. As brush was taken away, handsome granite and concrete gun platforms with brick parapets were discovered. Part of the original 1808 exterior brick wall was excavated, and weeds were removed from the old gun emplacements for a better viewing of early Fort Scammel.

Hilda Cushing, the Savior of Fort Scammel, died at the end of the bicentennial summer on October 26, 2008. Suffice it to say that Hilda was a unique individual. She is remembered for her kindness and generosity, as well as for being a tough, smart and determined lady.

An Epitaph for Fort Scammel

It seems as if all the forts and other public buildings constructed by General Thomas Casey have acquired a special ability for survival. In Washington, D.C., Casey built the Library of Congress and the State, War and Navy Building (now known as the Eisenhower Executive Office Building). He also took over the long-delayed construction of the Washington Monument, completing it in 1885. To quote from an article by journalist Rowena Henderson, "Silently, and with dignity, General Casey's buildings have stood through the decades, looking into the centuries beyond."

CHAPTER 3

THE KENNEBEC RIVER FORTS

THE ARTIST AT FORT POPHAM

The origin of Fort Popham goes back to the doomed Popham Expedition of 1607. In August of that year, one hundred English colonists landed on a windy point, half a mile from the mouth of the Kennebec River. In his classic book *Maine Forts*, Maine state librarian Henry Dunnack informs us that beginning with the construction of Fort St. George in 1607, six forts have been built on the Phippsburg peninsula.

The entrance to the Kennebec River remained strategically important until well into the twentieth

Cyrus Longley, the "Artist at Fort Popham." *Courtesy of Phippsburg Historical Society.*

century. Fort St. George's tenure, however, was short-lived. Following a severe winter that included a fire, discord among the colonists and the death of the colony's leader, George Popham, Fort St. George was abandoned, and the remaining settlers returned to England in 1608.

Moving on to the background of Fort Popham, Henry Dunnack tells us that the first building on the site known as Hunniwell's Point was a small fort erected in the eighteenth century to protect settlers from marauding French and Indians.

With the outbreak of the Civil War, Secretary of War Simon Cameron ordered an improved fortification to be built on Hunniwell's Point. General Joseph Totten, chief engineer of the United States Army from 1838 until his death in 1864, recommended the new fort be named after the leader of the long-forgotten Popham Colony. Work on Fort Popham commenced in November 1861, and one of the forty-odd soldiers involved with its construction was a young man from Bath with a talent for painting named Cyrus W. Longley.

A Third System Fort

America's Third System forts were developed following the War of 1812. Donna McKinnon gives us a description in her master's thesis "Portland Defended": "This system was implemented in stages from 1816 to shortly after the Civil War. Unlike the first two systems of harbor defense, the Third System began in relatively tranquil times. Earthen batteries, which needed constant repair, were ruled out in favor of brick and stone. Masonry fortifications also allowed for multiple tiers of casemated [fortified] gun emplacements."

Implementing Third System fortifications throughout the country was an expensive proposition, so it was not until 1858 that construction was begun on a Maine fortification, Fort Gorges in Portland Harbor. This was followed by Fort Popham a few years later. With the disruptions to the Union navy caused by the appearance of the Confederate ironclad *Merrimac* in 1862, Secretary of the Navy Gideon Wells became acutely aware of how vulnerable to attack coastal shipyards were.

The Kennebec River led not only to the shipping town of Bath, sixteen miles up the river, but also to the state capital of Augusta, thirty miles farther. As a result, what had been a leisurely project to guard the mouth of the Kennebec speeded up dramatically. Ironically, improvements in fortifications changed so rapidly during the Civil War that most of the forts begun at the start of the war would soon become obsolete.

Although it was never fully finished, Fort Popham was built of massive cut granite blocks shipped from Dix and Vinalhaven Islands in Penobscot Bay. Popham's location was important, occupying a rocky point where the Kennebec River broadens out into Atkins Bay, less than a mile from the sea. Henry Dunnack notes, "At this point the river channel is narrow and the current so swift, that any vessel attempting to pass must run close under the guns of the fort."

Robert Bradley writes in *The Forts of Maine, 1607–1945*, "As designed and largely built, Fort Popham is of a closed lunette form; that is, it is roughly crescent-shaped with defenses on all sides. In circumference the work measures 500 feet, while the sides facing the river rise to a height of over thirty feet. Enemy vessels attempting to sail up the Kennebec would have faced thirty-six, ten and twelve inch cannon arranged in two tiers of vaulted casemates."

Internally, Fort Popham was provided with a spacious parade ground containing two barrack blocks (which do not survive) a great subterranean cistern to afford a besieged garrison an ample supply of water for a long siege and four magazines to provide powder to the fort's maximum of forty-two guns.

Captain Henry W. Owen of Bath, who was stationed at nearby Fort Baldwin during the First World War, adds to Popham's description: "Had the fort been completed the main granite walls would have been surmounted by a deep earth parapet to protect the cannoneers from high angle fire, similar to the grassy parapets which grace so many other works of the period."

Captain Owen concludes:

> *The situation, armament and strength of the Fort were such that so long as its guns were well served it would have been practically impossible for warships of the Civil War period to have passed. A serious attempt to force the river entrance would have involved a landing and an attempt to carry the work from the rear, hence the moat, and small guns and loopholes commanding the beach, the moat and the land in the rear of the fort.*

The Soldier as Artist

What else do we know about the history of Fort Popham? It is from the aforementioned Private Cyrus W. Longley's drawings and watercolor paintings that we are given an insight into the soldier's daily life. Although he worked on the fort most days, in his spare time Longley sketched his comrades at work as they constructed the granite walls.

Cyrus Longley was born in Bath on January 10, 1841, the oldest of four children of Nathaniel and Sarah Longley. Cyrus was educated at schools in Bath, and as a young man he trained to be a photographer, a rare profession for the time. When he died in 1918, his obituary noted that he learned his trade at William Stearns's studio on Front Street. Records show that Longley joined the Seventh Company of the unassigned Maine Infantry Volunteers in 1863 at the age of twenty-two.

Young Longley served at Fort Popham from 1864 to 1865, where he worked with forty other volunteer soldiers from Maine on the construction of the fort. It is through Longley's primitive-style watercolor paintings that we have pictures of men laboring on a granite fort that was never completed. The composition of his paintings, however, reveals Longley's training as a photographer.

Longley did his drawing on sheets of drafting paper. One picture shows a long barrack building with several soldiers mustered outside. A large flag is flying on what was apparently the parade ground inside the fort. In other pictures, Longley shows his fellow soldiers working on the granite walls and fishing off a nearby pier. He also painted local ferryboats and fishing boats, the guardhouse, the parade ground, nearby lighthouses and a lively ice-skating party. Today, a twelve-foot mural of panels that Longley painted of Fort Popham and environs sits in the Phippsburg Historical Society Museum. It was presented by the Longley family in 1973.

The Phippsburg Historical Society has a newspaper article from 1892 in which another soldier, John Goff from Lewiston, describes a typical workday:

> *We half camped, half lived in shanties. It was get up early and eat a hearty breakfast. The food was better than most of the boys in blue got. Then at noon there was time for dinner and the smoking of a pipeful of our best tobacco before we went back to work again. We Lewiston fellows sort of cliqued together and it went hard with any other men who disturbed us.*

Roll call—from a watercolor by Cyrus Longley. *Courtesy of Phippsburg Historical Society.*

Entrance to Fort Popham during its construction. From a watercolor by Cyrus Longley. *Courtesy of Phippsburg Historical Society.*

Goff adds that "it was strange" no one was killed during the building of the fort. "The nearest that they ever came to killing a man was when one of the bosses was jammed between two pieces of granite so that he never did a day's work again."

Postwar Popham

After the war, work on Fort Popham came to a halt. For the next thirty years, the United States went through a difficult transition period, as memories of the Civil War remained painful. The soldiers in the Fort Popham garrison were demobilized, and the guns were scattered around the state. Two guns were sent to Bath. One old muzzleloader remains at Popham Beach and can be seen near the site of the former Riverside Hotel.

With the coming of the Spanish-American War in 1898, an attack by the Spanish fleet was feared. As a result, Fort Popham was once again garrisoned. Mines were readied for placement in the channel that led up the Kennebec River to Bath. Guns, eight inches in length, were mounted in anticipation of an attack by what turned out to be the vastly overrated Spanish navy. These were the last improvements made by the navy at Fort Popham. When the United States entered World War I in 1917, the entrance to the Kennebec River would again be defended, but not by an unfinished granite fort on Hunniwell's Point.

Fort Baldwin, built between 1905 and 1912, represents the War Department's response to the United States considering itself a global power following the Spanish-American War. This new fortification, built on a hill across the cove, was named in honor of Jeduthan Baldwin, a noted Revolutionary War engineer. Granite forts having been deemed obsolete, Fort Baldwin was built of reinforced concrete, the material now used in the construction of coastal forts. The fort consisted of three artillery batteries, each with a greater range and accuracy than the cannons of nearby Fort Popham. The gun emplacements were well hidden on a heavily wooded hillside that overlooked Popham Beach State Park and the Kennebec River.

Battery Coogan was named for Lieutenant Patrick Coogan from New Hampshire, who served in the Continental army. Battery Hardman was named for Captain John Hardman from Maryland, who also served in the Continental army. Hardman was taken prisoner at Camden, South Carolina, and died while a prisoner of war on September 1, 1780. Battery Hawley was named in honor of Major Joseph Hawley, who served in the

Civil War and died in 1905—the same year the battery construction was begun. This battery also housed the fort's original observation station and electronic equipment.

During World War I, Fort Baldwin and Fort Popham combined to house a garrison of two hundred soldiers. In addition to the three batteries, an electronically controlled minefield was installed to keep potential German raiders from advancing up the Kennebec. Following the war, Fort Baldwin's guns were removed, and on January 22, 1924, the State of Maine purchased Forts Baldwin and Popham for $6,600.

At the start of World War II, Fort Baldwin was reactivated. A five-story concrete fire control tower was built to maintain radio contact with other observation posts along the Maine coast. If an enemy vessel was spotted, observers in the tower could radio the enemy's position to soldiers manning the sixteen-inch guns on Peaks Island in Casco Bay.

It should be noted that most of the military personnel assigned to Fort Baldwin did not live at the fort but in houses around Atkins Bay. Gary Morong of the Phippsburg Historical Society told me the descendants of many of the families who were stationed at the fort have continued to live in the area.

An article by Doris Isaacson entitled "Last Coastal Defenders" describes the living conditions of the lonely soldiers who occupied Fort Baldwin, vacant since 1918, during the winter of 1941–1942:

> *That was some winter at the mouth of the Kennebec as Northeast gales swept across the Phippsburg peninsula. The wonderful people of Bath and Phippsburg made our primitive conditions more bearable. Cold sentries were warmed on many a midnight with hot drinks and pastries brought over the treacherous roads from Bath.*

Two Longleys

When he returned from the war in 1865, Cyrus W. Longley abandoned his job as a photographer and began what turned out to be a lengthy career as a civil servant for Sagadahoc County and the Town of Bath. His first employment was as a clerk in the registry of deeds in Bath, where he worked for Register Henry Bovey. Several years later, he moved on to work in the office of Clerk of Courts Joseph Hayes. Longley was next elected register of the Sagadahoc Probate Court in 1872. With the exception of an eight-year period, he held the office until his death in 1918.

Longley was also clerk of the Board of Common Council for twenty-five years, and for five years he was the clerk of Bath. In addition, Longley was a justice of the peace for many years, beginning in 1873. Reportedly, he was very proud of his collection of the signatures of the numerous governors under whom he served.

On the occasion of his seventy-third birthday (January 10, 1914), the *Bath Independent* cited "his remarkable record." It continued, "Mr. Longley is one of the oldest active county officials in the State and his reputation is not confined to Maine. He is considered one of the best officials in probate matters in the nation." Longley was also a former chancellor of the Acadia Lodge and keeper of records and seals for the Knights of Pythias.

Longley died of bronchial pneumonia in 1918 at the age of seventy-seven. His obituary described him as "the veteran incumbent of the court house having held an office almost continuously since it was built." Longley's wife, Hannah, predeceased him, and a son and a daughter survived him. Longley left his son, Albert, his Civil War sketchpad, which included his drawings and paintings of Fort Popham.

Cyrus Longley was a highly respected member of the Bath community, as illustrated by this excerpt from the *Bath Daily Times*:

> *Never has Sagadahoc County had a more faithful or efficient official than Mr. Longley and it is doubtful if any man has had a wider or longer acquaintance with the citizens of the county than Mr. Longley. He was everybody's friend and those who went to the courthouse to appear before the probate court for the first time were always shown the most kindly courtesy and consideration by him.*

Cyrus Longley's great-granddaughter, Diane Longley, is a retired Phippsburg schoolteacher who lives in Woolwich, Maine. Diane has led an active retirement. She published her first book, *Steel over the Kennebec*, in 1978, and over the years, she has written numerous historical articles for the *Bath Times Record*. Diane recently completed her sixth book, *A History of Woolwich*.

As a child, Diane remembers visiting her grandmother on Court Street in Bath and being shown the sketchbook and mural of Fort Popham created by her great-grandfather. Diane published her second book in 1986, *Fort Popham, Maine, A Civil War Fort*, which included many of her great-grandfather's sketches. Subsequently, Diane has given lectures on Longley's artwork and shown slides of his paintings that are on display at the Phippsburg Historical Society.

For many years, the Longley family treasured Cyrus Longley's sketchbook and watercolor paintings, never dreaming that another set existed. In the early 1980s, Earle Shettleworth, director of the Maine Historic Preservation Commission, discovered a nine-panel set of watercolors of Fort Popham in a Yarmouth antique shop. They were good primitive paintings, so he purchased them for the state museum. The only clue as to the artist were the initials: CWL.

The mystery was solved in 1986 when Sheila McDonald, from the Bureau of Parks, ran across Diane Longley's book on Fort Popham, which contained reproductions of the pictures done by her great-grandfather. Apparently, Cyrus Longley had made a copy of his earlier paintings— possibly for someone who served with him at the fort? Of course, this raises the question of whether Longley did any other artwork. To date, nothing has been discovered.

Afterword

Fort Popham was again in the news on September 22, 1995, when the *Portland Press Herald* ran a story declaring that a mine at the entrance to the fort had been declared "harmless." Although the old harbor mine dated from the early 1900s, there were fears that it still might be "live." The weathered steel ball was four feet in diameter and weighed an estimated eight hundred pounds.

Greg Lippman, an army explosives expert, was the first to raise a concern when he visited the fort in August 1995 as part of a nationwide search for unexploded bombs at old military installations. Local residents took the matter seriously, even though kids had been playing on the mine (and throwing rocks at it) for years.

An explosives team from Rhode Island was summoned, and after drilling several holes into the mine, Lieutenant Commander Robert Wiegard announced, "It's hollow. We saw right into the bottom. We didn't think it was live, but there was always a chance." He continued, "The Navy keeps records on mines that have been disarmed, but there were no records on this one."

"Supporters Hope They Can Hold the Fort" read the headline in the *Maine Sunday Telegram* on February 19, 2006. The concern expressed in the article was that "decay and deterioration was weakening Fort Popham's construction." An engineering firm was engaged, and the Friends of

An inactive mine at the entrance to Fort Popham. *Author's collection.*

Phippsburg joined with the state to raise money to repair the stonework and brickwork. The project was completed in 2008.

Fort Popham has been a Maine State Historic Site since 1926. Admission is free, and visitors are permitted to walk around the galleries.

An aerial view of Fort Popham today. *Courtesy of Friends of Fort Popham.*

A series of excellent interpretive panels is mounted inside the walls of the gallery facing the sea. The park ranger, Brian Murray, told me recently, "All things considered, the fort is in pretty good shape. The stones haven't moved in 150 years."

THE SONS OF FORT WESTERN

It is unlikely that the nineteenth-century German philosopher Karl Marx was familiar with the history of Fort Western in Augusta, Maine, when he developed his famous theory that economic forces determine the course of history. Whether one accepts this interpretation of history or not, there is no denying the fact that although Fort Western was fortified, its chief function was mercantile.

Fort Western and Fort Halifax, its sister fortification to the north, were each built in 1754 as part of a defense program instituted by the Massachusetts

Fort Western was built on the banks of the Kennebec River in 1754. *Collections of Maine Historical Society, Courtesy of Maine Memory Network.*

Bay Colony to defend the local population against raids from Canada at the start of the French and Indian War. Like a number of the forts discussed in this book, Fort Western was never attacked. Instead, the fort acted as a supply depot for Fort Halifax, the real guardian of the region, sixteen miles farther up the Kennebec River.

Beginning with Captain James Howard in 1754 and continuing with his sons Samuel and William, the story of Fort Western becomes primarily that of a commercial operation.

Seventeenth-Century Background

George Dow, in his booklet *Fort Western on the Kennebec*, reminds us that "long before the coming of the white man, the river had not only been a center for several tribes of Indians but also a highway for communication (and trade) between the interior of Maine and the sea coast."

It was therefore not surprising that the area around the site of Fort Western would grow from the impetus of trade. Several years after their arrival in 1620, the Plymouth Pilgrims began to fan out and establish trading posts. As early as 1625, a group of Pilgrims sailed up the Kennebec River as far as present-day Augusta, seeking to exchange goods with the local tribes (Abenaki) living in the area.

That year, Plymouth governor William Bradford wrote, "After harvest this year [when the Pilgrims first had a surplus of corn] they sent out a boatload of corn 40 or 50 leagues to the eastward, up a river called Kennebec. They laid a little deck over her amidships to keep the corn dry. They brought home 700 pounds of beaver, besides some other furs."

Several years later, the first trading post was built close to the future site of Fort Western, where corn, beads and other goods were swapped for furs trapped by the Abenaki. The result was that the Pilgrim traders were able to generate enough profit to eventually buy out their creditors in London. Governor Bradford described the situation: "They erected a house up the river in the most convenient place for trade and furnished the same with commodities for that end, both winter and summer."

By 1628, a permanent trading post was built at Cushnoc (an Abenaki word meaning "a place on a tidal river where the current overcomes the tide") near present-day Augusta. A second post was built upstream at Winslow, the future site of Fort Halifax, and a third was built at the mouth of the Kennebec. At the same time, Pilgrim merchants began to trade wampum beads from Long Island, which soon became a popular item among the Kennebec tribes. William Bradford adds, "Afterwards the inland peoples could scarce ever get enough of it."

In the five-year period from 1631 to 1636, considerable prosperity was achieved, enabling Pilgrim merchants to ship 12,150 pounds of beaver and 1,156 pounds of otter to England. As previously noted, it was the fur trade on the Kennebec River that relieved the Pilgrims of their financial difficulties. According to George Dow, "This extricated them from the clutches of the Merchant Adventurers of London."

Increased profits led to more competition, however, causing the price of fur to drop by the 1640s. Trade with the Kennebec tribes declined further when many of the Abenaki fur trappers were wiped out by pandemics like smallpox and spotted fever, which were brought by early colonists. Over time, relations between the Abenaki and English settlers worsened, and many of the tribes allied with the French in Quebec.

The result was a series of conflicts with Indian tribes that raged across New England for the next seventy-five years, causing English settlements in the Kennebec Valley to disappear. Indeed, following the particularly vicious King Philip's War (1675 to 1676), most of the Province of Maine was abandoned.

Historians consider King Philip's War, when bloody raids swept as far south as Massachusetts and Rhode Island, to be the greatest calamity to occur in seventeenth-century New England. Dozens of towns were destroyed, local economies were ruined and the population was reduced by one-third.

Proportionately, it was one of the bloodiest and costliest wars in the history of North America. The Indian population also suffered greatly. It is estimated that the Abenaki tribes in western Maine were reduced by 60 percent. In the Kennebec region, approximately forty warriors remained out of an original population of three thousand.

As a result, the Province of Maine remained a virtual no-man's land until well into the eighteenth century. Towns gradually reappeared along the coast, although it was not until mid-century that settlers began moving into the Kennebec Valley.

Two Forts Are Built

Fort Western was most likely named for Thomas Western, an English friend of Massachusetts Bay governor William Shirley. Shirley was interested in extending the presence of his colony at the conclusion of King George's War in 1748. Governor Shirley was on the peace commission at the war's end, and he returned frustrated by Britain's failure to secure lands in northern and eastern Maine. Building a line of forts on the Kennebec River seemed like a solution to the ongoing struggle with the French and their Indian allies. All he needed was an excuse.

When William Lithgow, a local "truckmaster" (an individual who supervised a "truckhouse," or trading post), reported that the French were building a settlement at the Great Carrying Place at the headwaters of the Kennebec River, Shirley seized the opportunity to propose construction of military bases farther up the river. The Massachusetts General Court (colonial legislature), however, balked at the expense. Shirley therefore struck a deal with the newly formed proprietors of the Kennebec Purchase to build Fort Western as a supply base for an advanced post, Fort Halifax, eighteen miles farther up the river.

Fort Halifax's location below the Teconnet Falls, opposite present-day Waterville, was of strategic importance as it was on a major French and Indian inland travel route. Its primary function was to serve as an outpost to protect the new settlements farther to the south. When a shipload of French Huguenots landed in 1751, they were encouraged to settle in the Kennebec Valley, specifically at Cushnoc. Because of unstable

Massachusetts governor William Shirley was responsible for having Fort Western built to serve as a supply base for Fort Halifax, located farther up the Kennebec River. *Collections of Maine Historical Society.*

relations with the Abenaki, however, the immigrants' arrival was delayed for several years.

Construction on both forts began in the summer of 1754. Fort Halifax, however, was never completed. The garrison's commander, William Lithgow, had chosen a vulnerable location, causing the fort's design to be changed and an additional structure built on a neighboring hill. At the end of the French and Indian War, Lithgow resumed his job as a truckmaster, trading with local tribes in the otherwise abandoned fort.

Governor Shirley had proposed that two forts were needed on the upper part of the Kennebec River. Gershom Flagg, a builder from Boston, was engaged to supervise the construction of Fort Western at the Cushnoc site. Fearing an attack from the north, Flagg had timbers for both forts cut downriver and dragged back upstream.

Many of the eight hundred men recruited to build Fort Halifax were detained for several weeks to help erect blockhouses and barracks at Fort Western. An article in the *Boston Gazette* provided a description: "The fort is about thirty feet high and commands the river. We have raised two blockhouses and two watch boxes. Pickets (palisades) are cut and a fine road is made from the water up the bank."

James Howard was a Scotsman who had been living in the St. George (Thomaston) area of Maine since the 1730s. He served in King George's War, where he met William Lithgow. When construction on the Kennebec forts began in 1754, James Howard was put in charge of the garrison at Fort Western. During the French and Indian War, however, Howard was under the command of Colonel William Lithgow at Fort Halifax.

Although Howard was paid as a lieutenant, his men called him "captain." In a letter to Governor Shirley, Howard asks for more guns and states, "We have no colors [flags]." At the same time, at Fort Halifax, Lithgow was complaining about a lack of supplies.

The next year, Howard pleaded for more men: "It is very probable we shall have some of our French and Indian enemies to visit us this spring and our number of men is small and the ground around our fort is advantageous for our enemies." Hostile Abenaki continued to lurk nearby, and there are several accounts of skirmishes that took place outside the fort: "One man hath a bullet lodged in his leage [*sic*] and is wounded in several parts of his body."

The Kennebec Valley was considered a backwater area during the French and Indian War, so clearly Howard's and Lithgow's letters fell on deaf ears. An order dated July 15, 1755, further reduced the joint garrisons of both forts to eighty men, only twenty of whom were stationed at Fort Western. Soldiers were needed elsewhere, especially to support General Wolf's attack on Quebec.

The *Interpretive Guide to Fort Western* emphasizes that throughout the French and Indian War, Fort Western's mission remained that of a supply post. The

The fort's storehouse was also used as a barracks for soldiers and living quarters for the Howard family. *Courtesy of Old Fort Western.*

guide continues, "It was a fortified storehouse and never a military post. Fort Western was built by the Kennebec Proprietors strictly for commercial purposes." Perhaps this is why James Howard's requests for reinforcements fell on deaf ears.

Although Fort Western was never attacked, the same could not be said for the garrison at Fort Halifax. The first attack came in the fall of 1754, when a force of one hundred Indians attacked a small detachment of soldiers who were cutting wood. They were driven off when troopers arrived from the fort. Over the next few years, shots were exchanged between soldiers and smaller bands of Abenaki on a number of occasions.

The Howards at Fort Western

It was not until the fall of Quebec in 1759 that it was considered safe for civilians to move into the Kennebec area. As the threat of war declined, the population along the river finally began to increase. Fort Western's garrison was reduced, and in 1767, the fort was decommissioned. That same year "Captain" James Howard purchased the lands and buildings of Fort Western from the Kennebec Proprietors for £270.

By the end of the French and Indian War, James Howard was the largest landowner in the Fort Western area. The Howards, father and sons, also owned the schooner *Western*, which began to trade on the Kennebec as early as 1759. There is a reference in Dow's book to the Howards as having "a large stock of cattle, several sloops and carrying on considerable trade" in 1765.

To quote again from the *Interpretative Guide*, "Rather than going home at the end of the military period (1754–1760), the Howards decided they were home." James Howard would run a variety of mercantile operations out of Fort Western for many years, and members of his family would reside in the three-story main building of the fort until the middle of the nineteenth century. The *Interpretative Guide* informs us that the Howard family helped make the settlement of Augusta possible because of the store they operated at Fort Western.

There were four Howard sons and a daughter, Margaret. We know little about the two older boys except that the oldest, John, succumbed to mental illness. James Howard's youngest son, William, was the one who took charge when his father turned the fort over to his sons. William was fourteen when his father brought him to Fort Western in 1754. By the time he married in 1770, the young man had become a skilled blacksmith, as well as manager of the family store.

Samuel Howard, another son, moved to Boston in 1774. By then, he and his brother William had formed a partnership, S&W Howard. Samuel ran the family's business operations in Boston. He purchased goods and shipped them to Maine, where William sold them to the Kennebec settlers who were dependent on the outside world for both necessities and luxury goods. George Dow adds, "Howard sloops carried staves, shingles, salmon, moose skins and furs to Boston and returned filled with pork, corn, flour, shoes and articles of clothing." Howard ships reportedly sailed as far as Newfoundland and the West Indies.

William Howard spent his adult life at Fort Western, where he operated what became a multipurpose store out of the main building. The original barracks building was divided in thirds, with the store in the middle. William's residence was at one end and the blacksmith's forge at the other. Later, when the forge was converted into a residence for Captain James Howard's family, it was moved to another building on the property.

This bridge was built across the Kennebec River in 1797. It connected Fort Western with Augusta, spurring commercial development. *Courtesy of Old Fort Western.*

The Howards' store served as a bank, a tavern, a lawyer's office and a surveyor's office. Fort Western's program coordinator, Judy Sample, calls it "an eighteenth-century Wal-Mart, with a little bit of everything." According to Ms. Sample, William Howard was the brother who ran the Howard commercial empire. "He was the guy who made it all work."

This is not to minimize the contributions of "Captain Samuel," as he was known. It was Samuel who made the contacts for Howard operations in Boston and beyond. Samuel also got into the transportation business. Regular sloop service was set up between Kennebec and Boston in 1790 during the ice-free months.

The Revolution and Beyond

Fort Western was involved in the American Revolution, albeit peripherally, on two occasions. The first occurred in the fall of 1775, when the fort was used as a jumping-off point for Benedict Arnold's invasion of Canada. Arnold ferried one thousand men and supplies from Newburyport, Massachusetts, up the Kennebec River to Gardiner, six miles below Fort Western. At that point, the army transferred to bateaux, river craft especially designed to navigate the upper reaches of Maine's rivers.

This is not the place for an analysis of the Arnold Expedition. Suffice it to say, Arnold proceeded to Fort Western, where he stopped for a few days to consolidate his forces and repair some of the poorly constructed bateaux. His army included a number of famous names, including renowned rifleman Captain Daniel Morgan; Aaron Burr, a future vice president; and Captain Henry Dearborn, destined to be secretary of war under Thomas Jefferson.

The Howards welcomed Arnold's army. George Dow writes, "Colonel Arnold and some of his officers were entertained by Squire Howard at the 'Great House.' Other officers were lodged at the Fort where they were exceedingly well entertained." The common soldier, however, had to fend for himself. Private Caleb Haskell from Deer Isle wrote, "Several of the companies have no tents so encamped on the ground. We are very uncomfortable it being very rainy and cold."

Proceeding upriver from Fort Western, Arnold's invasion force was plagued by disease, desertion and miserable weather. When the remaining six hundred exhausted and starving men finally attacked Quebec on New Year's Eve 1775, they were routed by the French in a driving snowstorm.

We can only speculate as to whether Arnold questioned the Howards about the Kennebec route to Canada. Did they suggest that it might be more difficult than Arnold imagined? And how did the Howards react when deserters from the march reappeared, having taken much of the army's food?

Four years later, in the summer of 1779, Fort Western was again a rendezvous point. This time it was for those men retreating from the ill-managed Penobscot Expedition. Paul Revere, commander of artillery, was a member of the amphibious force that the Massachusetts General Court sent north to dislodge the British at Castine in Penobscot Bay. The arrival of a British fleet caused the American ships to flee up the Penobscot River, where most were burned. The survivors of the disaster, including Paul Revere, stopped at Fort Western for a few days before proceeding to Boston.

During the Revolution, James Howard was appointed head of the local Committee of Safety. His sons, Samuel and William, both saw active duty, especially William, who served as a lieutenant colonel in the Penobscot Expedition. Because of his familiarity with inland Maine, William was able to guide the remnants of the defeated American army through the woods to safety. After the Revolution, William returned to Fort Western and managed the family business. Account books indicate that he operated the store until 1804 and possibly until his death in 1810.

The family patriarch, James Howard, spent his last years at home in the three-story barracks/storehouse surrounded by his second wife and a brood of young children. He died on May 15, 1787 at the age of eighty-six. Martha

The Old Fort Western complex today. The fort is open to the public during the summer and runs educational programs throughout the year. *Courtesy of Old Fort Western.*

Ballard reported his death in her diary: "Judge Howard departed this Life; a Suden [*sic*] Change, he was well and Dead in about three hours. A larg [*sic*] funeral."

On occasion, Samuel Howard returned from Boston with his family to escape the diseases of the city and enjoy the cooler climes of Maine. Samuel died at Fort Western on March 30, 1799. Howard descendants continued to live in the main building until at least 1840. Records show that it finally passed out of the family in 1866.

In 1919, Guy Gannett, the great-great-grandson of James Howard, rescued the now derelict main building from destruction. Gannett then proceeded to restore Fort Western at considerable personal expense, presenting it to the City of Augusta in 1922.

Old Fort Western is open to the public from Memorial Day to Labor Day. Educational programs are run throughout the year.

CHAPTER 4

THE LINCOLN COUNTY FORTS

THE SCRIBE OF FORT EDGECOMB

Moses Davis was twenty-seven years old in 1770, when he moved from Newburyport, Massachusetts, to Edgecomb in the Province of Maine, across the Sheepscot River from Wiscasset. A pious and hardworking young man, Davis soon became a respected member of the Edgecomb community. He was a justice of the peace for many years, and in 1775, he was chosen to represent the recently incorporated (1774) town in the Massachusetts Provincial Congress at the start of the American Revolution.

After the Revolution, Moses Davis represented Edgecomb at the Massachusetts Convention in 1781, which led to the Articles of Confederation. Later in the decade, he was a delegate at the Massachusetts Convention in Boston, where he voted for the ratification of the Constitution of the United States. As we shall see, Moses Davis (as a much older man) was also directly involved in the building of Fort Edgecomb.

In the early nineteenth century, with tensions increasing between Great Britain and the United States, the federal government decided to build a string of forts to defend America's coastline against the powerful British navy. The fortifications, known as the Second System, were begun in 1807. The Second System was an improvement over the First System in that it introduced a better design to protect gun crews from enemy fire. It also recommended the replacement of foreign engineers with American ones, many of whom were recent graduates of the newly established United States Military Academy.

The immediate background to deteriorating relations between the United States and Great Britain was the 1807 attack by the British frigate *Leopard* on the newly outfitted American frigate *Chesapeake* off the Virginia Capes. Although four British deserters were found among the crew of the battered *Chesapeake*, the American public was outraged and demanded a military response.

Rather than go to war (and in an attempt to stop attacks on American merchant vessels), President Thomas Jefferson and Congress passed the Embargo Act in December 1807. The embargo blocked trade with England and France, locked in the Napoleonic Wars, until those countries agreed to respect American neutrality. The new law virtually crippled the mercantile economy of New England, although it did not substantially damage the economies of England and France.

After issuing the Embargo Act, Congress then appropriated $1 million to begin construction on the previously described Second System of fortifications. Several forts were begun along Maine's coastline, which was considered particularly vulnerable.

Fort Edgecomb was built in 1808 and is located directly across the Sheepscot River from Wiscasset, which was considered a vital port on the Maine coast. In the late eighteenth century, Wiscasset was second only to Portland in terms of shipping tonnage from Maine. By 1800, the town had ten piers on the Sheepscot River. The town's primary export was timber for Europe and the West Indies, but there was also a fleet of fishing and merchant vessels that traded up and down the coast. In his book *Battery and Blockhouse: A History of Fort Edgecomb*, Joshua Smith writes:

> *Wiscasset's harbor was so good that it attracted the attention of the federal government who considered the port as a potential Navy Yard in 1802. The harbor was deep and could safely shelter large fleets and was safe from storms…but there were disadvantages too. Maine's thick fogs, lasting as long as fifteen days, concerned the Navy, as did the isolation of Wiscasset from major population centers.*

Congress put Secretary of War Henry Dearborn in charge of building Maine's fortifications. Dearborn was from New Hampshire and had fought in the American Revolution. After the Revolution, he lived in Gardiner on the Kennebec River and had represented the Province of Maine in Congress from 1793 to 1797.

The view of the Sheepscot River from Fort Edgecomb. *Author's collection.*

Earthworks along the Sheepscot River provided protection for the nearby town of Wiscasset. *Author's collection.*

The Five Moseses

Since arriving in Maine with his wife, Sarah, Moses Davis had become a leading citizen of the town of Edgecomb. *The Early History of Edgecomb Maine*, compiled by L.I. Davis in 1986, describes his public life:

> *The first permanent settlement on Davis Island was made by Moses Davis, Esq. in 1770. Moses and Sarah Davis had six children born to them on Davis Island. Along with clearing the land and establishing a farm, Moses Davis followed the carpenter's trade. As a Justice of the Peace from 1776 until his death in 1824, he married one hundred and eighty three couples along with performing many types of legal matters. At the first town meeting he was elected town Clerk, he later served as Treasurer, and for nineteen years as a Selectman.*

Moses Davis was sixty-five years old when he heard of the plans to build a fort in the Wiscasset area. In addition to his town duties, Davis had also become a confirmed diarist. It is largely through his writings that we know as much as we do about the history of the town and the construction of Fort Edgecomb.

According to Joshua Smith, Secretary of War Dearborn ordered a senior army engineer, Major Moses Porter, to oversee the building of "batteries from the mouth of the Kennebec as far east as Castine harbor." Porter had served under Washington and Henry Knox during the Revolution, and his knowledge of artillery made him well suited to supervise the construction of Maine's mid-coast forts.

In Wiscasset, Porter was further instructed to consult with "leading men as to where

Moses Porter, the builder of Fort Edgecomb.
Courtesy of Friends of Fort Edgecomb.

to put the battery." Porter turned to a local businessman and ship owner, Moses Carlton, to handle the details of the job. Carlton was responsible for finding contractors, arranging for the delivery of supplies and paying the bills.

On May 21, 1808, Moses Davis wrote in his diary that Porter and Carlton came to Davis Island "to look for a place to make a fort." (Davis Island is directly across the river from Wiscasset.) The upshot of the meeting was that on May 30, 1808, the United States government purchased three acres on Davis Island from Moses Davis for $300.

Major Porter wasted no time getting started. The very next day, he began to set out boundary stakes for the fort. At the same time, he moved in with the Moses Davis family, with whom he would live for the next six months. Once construction began, two other men named Moses appeared. One was Moses Davis Jr. The other, Moses Dodge, was a hired hand who worked for Moses Davis Sr. and, as Joshua Smith notes, "assisted in the fort's construction." Meanwhile, Moses Porter traveled back and forth between his various jobs, overseeing the construction of four other coastal forts, or batteries, that had been authorized by Congress.

Moses Davis, his son and Moses Dodge spent the next few weeks clearing the site of trees and brush and hauling stones in preparation for building the fort. By midsummer, the senior Davis had withdrawn from the heavy construction work, leaving it to his son, Moses Davis, and several other laborers. Joshua Smith says, "Less is known about these men and other laborers who built the fort, although we know that John Dodge, of Wiscasset, was one of the masons who worked there in 1809."

Although Moses Davis's prose was rather terse, his details about the fort's construction are invaluable. On August 2, 1808, Secretary of War Henry Dearborn arrived to inspect the progress. We know that Davis did not approve of Dearborn because he supported President Jefferson and the unpopular Embargo Act. Nevertheless, he recorded, "Wind southerly, boys finished mowing grass and got in the hay. Secretary of War came here to the battery and gave directions how to proceed." Interestingly, Davis continued work on the fort at the same time he was petitioning the Massachusetts General Court to "assist us in redress of the Embargo laws."

After Secretary Dearborn's departure, the pace of construction accelerated. Moses Davis Jr. began to haul planks for the gun platforms, while Moses Dodge and others brought rocks, brick, timber and sod for the battery. By the end of the fall of 1808, the blockhouse (which served as a shooting platform, living quarters and a storehouse) was nearly finished, causing Moses Porter to anticipate that the fort would soon be ready for troops. Moses Davis disagreed that the fort was nearing completion. In his description of the building process, Joshua Smith describes Davis's views: "Moses Davis's diary entries from 1808–1809 make it clear that the greatest labor in constructing Fort Edgecomb was building the stone and brick battery where the cannon would be positioned. The wooden blockhouse… was in some ways almost an afterthought."

The two men clearly had different priorities. Moses Davis, watching from a distance, was more concerned that the battery and the powder magazine be completed. At such point, the fort would be able to function against an enemy attack. In the short run, Davis probably felt the garrison could exist without a permanent shelter for its defenders.

Moses Porter's priority had a supporter in the person of Moses Carlton. The Wiscasset businessman and ship owner wrote to Secretary of War Dearborn in 1808, "The blockhouse is nearly finished and is the best piece of work I ever saw. Major Porter pays every attention to his business and has everything done in the best matter possible."

Work on the fort proceeded through the winter of 1808 and 1809 until it was completed in time to celebrate incoming President James Madison's inauguration on March 4, 1809. Moses Davis noted the event in his diary: "They fired seventeen guns at the battery today to welcome Madison into the Presidency." Interestingly, despite the coming war with England, there were only two times when Fort Edgecomb's cannons would be officially fired. (The second time was February 14, 1815, to signal the end of the war with Great Britain.) To this, Joshua Smith adds, "The gunfire celebrated James Madison's inauguration, but it was also the first time these cannon fired; they had only been mounted on their carriages a few days previous. The salute announced to all within earshot that this new military post was operative and its garrison on duty."

Robert Bradley describes the various components of the completed Fort Edgecomb in his book, *The Forts of Maine, 1607–1945*:

From the river bank to high ground Fort Edgecomb consisted of a massive stone revetment with twin bastions supporting two eighteen-pounders. A brick magazine [for ammunition] was buried beneath the eastern bastion and was fitted with a passage ten feet long…Above the lower batteries was a crescent-shaped earthwork which protected a single gun, a fifty pound Columbiad. Both of these levels were protected in the rear by a palisade. Behind the upper gun emplacements was a wooden blockhouse.

Visitors today can see the impressive blockhouse and the various levels of earthworks. Bradley describes the blockhouse as "a remarkable post and beam structure of octagonal plan with an overhanging second story." He adds, "In height it rises thirty-four feet to the top of a watch box. The blockhouse was originally equipped with two carronades."

Joshua Smith tells us that, given a lack of documentation, it is difficult to know the exact details of the fort's armament. For example, it is unclear whether there were two or four cannons facing the river from the revetment and whether they were eighteen- or twenty-four-pounders. There also may have been more than one of the fifty-pound Columbiad guns.

Despite a lack of enthusiasm for the policies of the government, particularly the Embargo Act, the Davis family worked hard to finish the job in time to celebrate Madison's inauguration. On February 23, 1809, hired hand Moses Dodge had his team of oxen haul cannons to the fort, where they were quickly mounted on their carriages.

One can imagine their happiness when news arrived that outgoing President Jefferson had repealed the Embargo three days before he left office. Even Moses Davis was pleased when he wrote, "Seven or eight vessels went down river for Europe and other ports."

The Road to War

Having completed his work at the fort, Major Moses Porter left his post, and Captain John Binney took command of Fort Edgecomb in March 1809. Upon hearing that the Embargo had been repealed, Binney ordered the fort's twenty-four-pound cannon to fire a salute. In a letter to his wife, Binney wrote, "This town is in an uproar...every man, woman boy, girl, horse, dog, cat, pig, duck and all living things are rejoicing. You can not imagine anything more noisy."

Between 1809 and the declaration of war in 1812, Fort Edgecomb was occupied by a garrison of between forty and forty-five men. Most of the enlisted men had been recruited from the area and had signed up for the required five years, having nothing better to do. Binney, a young artillery officer, had his work cut out for him. None of his officers had any previous military experience, which meant the men under them suffered from a lack of professional direction. Life for the enlisted man at Fort Edgecomb was primitive. Discipline was harsh, and not surprisingly, some men deserted.

Moses Davis mentions two desertions in his journal. On May 30, 1809, he wrote, "Edward Blanchard ran away from the fort," and on Saturday, August 5, 1809, he wrote, "Corporal Payson ran away from the garrison this evening." Although we don't know the specific punishments handed down at Fort Edgecomb, the army treated desertion severely. This could include a court-martial, flogging and, frequently, confinement in a dark chamber somewhere in the recesses of the fort.

Davis apparently hired off-duty soldiers to work on his farm. Although this practice was prohibited by the War Department, it was frequently ignored, which had benefits for the military. Davis recorded that his son hauled manure and plowed "a cabbage and turnip yard" for a garden near the fort. Davis also sold produce from his farm to the soldiers and even extended them credit, which Joshua Smith says "was probably more than most merchants were willing to do."

Smith's summary of the situation on the eve of war is worth noting:

> The overall impression of life at Fort Edgecomb is that it was not an especially happy existence. Officers squabbled, the sergeants beat the privates, the food was inadequate, the duty monotonous…but everything was about to change as the nation slid into war. On June 23, 1812, news arrived that Congress had declared war against Great Britain.

The War of 1812 and Aftermath

With the outbreak of war, life at Fort Edgecomb changed. The fort served as a training center for military expeditions and, as the war progressed, a detention center for British prisoners. Captain Binney and an expanded company of recruits left in June 1813 with orders to join a regiment in upstate New York. Captain Benjamin Poland's Thirty-fourth Infantry Company replaced him, although they were soon ordered to the Canadian border. Joshua Smith adds, "In addition to serving as a depot for recruits the post had another role as well, that of a defensive battery designed to protect American shipping and Wiscasset harbor."

In the summer of 1813, sixty British prisoners were assigned to Fort Edgecomb. This concerned the officer in charge, Second Lieutenant Calvin Croker, who had only a handful of "invalid soldiers" with which to guard them. Additional prisoners were sent from jails in Boston, since the antiwar Massachusetts legislature had passed a measure prohibiting the use of local jails to house British POWs. Given the lack of a garrison, escape was easy, and many POWs slipped away to the Maritime Provinces.

Fort Edgecomb continued to be garrisoned irregularly as troops were shipped off to where they were most needed. Until the spring of 1814, the Maine coast was relatively quiet. Then, in June 1814, HMS *Bulwark*, a seventy-four-gun ship of the line, appeared off Seguin Island below the Kennebec River. The huge ship posed a particular threat to the coast because of the smaller boats on its decks, which could sail up rivers and attack vulnerable ports. Moses Davis reported what happened next:

> *June 10, 1814: An English 74 and a frigate came to the mouth of the river and sent her barges up as far as Fowles Point* [on Westport Island], *drove away militia and took two six pounders and sank them in the river. After they went away with their barges our people got the cannons and hid them in the woods. A number of Militia came from Wiscasset to this fort to help guard it.*

Wiscasset went on high alert. On June 22, a local citizen filed this report to a Boston newspaper:

> *They approached to within a few miles of the fort, opposite this town, with the avowed intention of coming to the wharves and burning the shipping; but hearing our alarm guns and ringing of the bells, judged that we were prepared for them and retreated to their ships at the mouth of the river after robbing a few houses.*

At the end of a long and anxious summer, the British suddenly appeared in Penobscot Bay. In less than a month they raided and looted Hamden and Bangor. Nearby Castine would remain under British control for the rest of the war. Additionally, the British captured or destroyed a considerable amount of American shipping as far down east as Machias.

When he heard about the Penobscot Bay attack, Moses Davis recorded: "The militia came from Wiscasset this afternoon in order to guard against the enemy who are reported coming with an intent to destroy this fort and Wiscasset." Militia went back and forth across the Sheepscot River throughout the month, since it was by no means clear what the British intentions were. "The like I never saw," wrote Davis. At one point he packed away a valuable clock for fear the enemy would take it.

By late fall, the threat of a British attack was over, and on February 14, 1815, a happy Davis wrote, "News came of Peace betwixt Britain and

The Fort Edgecomb blockhouse today. *Author's collection.*

America, which made a great rejoicing and firing cannon at Wiscasset and at the fort." Indeed, another round of celebrating occurred when it was announced that the Senate had ratified the treaty. Crowd control was difficult under the circumstances. Davis complained in his journal that he had to send Moses Dodge "after our canoe the soldiers stole."

A small garrison remained at Fort Edgecomb until 1816. During the Civil War, volunteers briefly manned the fort in 1864, when the Confederate raider CSS *Tallahassee* plundered and destroyed numerous vessels in the Gulf of Maine.

Moses Davis, the indefatigable Scribe of Fort Edgecombe, died on November 18, 1824, at the age of eighty-one.

Fort Edgecomb benefitted from local preservation efforts in the last quarter of the nineteenth century. This was followed by Governor Percival Baxter's purchase of the fort and surrounding land from the federal government in 1923 for $501. Today, the fort is a State Historic Site. The Friends of Fort Edgecomb was formed in 1993 to help maintain and promote interest in the fort.

PIRACY AND TREACHERY ON THE PEMAQUID

For hundreds of years, Pemaquid, an Algonquian word meaning "point of land," had been the site of Indian settlements. By 1617, however, the last native village had been destroyed or abandoned. Beginning about 1610, Europeans established seasonal fishing stations, and by the mid-1620s, English settlers had begun to move into the area. In 1630, the first of four forts was built on a plot of land that overlooked the mouth of the Pemaquid River, not far from present-day Damariscotta.

Abraham Shurte, an agent acting for merchants from Bristol, England, erected the first "fort," which was essentially a fortified warehouse constructed to protect valuables from theft. Shurte was also directed to build a trading post at the newly established English settlement on the peninsula. The strength of Abraham Shurte's fort was soon tested. Within two years, Dixey Bull, the first pirate to pillage the New England coast, would attack it.

Who was Dixey Bull? His real name is unknown, and as we look back over a span of four centuries, the facts are often obscured by legend. At a time

View of John's Bay from the bastion of Fort William Henry. *Author's collection.*

when most pirates in the seventeenth century were pursuing Spanish gold in the warm waters of the West Indies, why was Dixey Bull operating in the Gulf of Maine?

The story begins innocently enough. Dixey Bull may have been an English sea captain who was trading beaver pelts and other goods with Indians along the Maine coast. In June 1632, he was in Castine Harbor when a French force attacked the Plymouth Colony trading post. Bull's ship was seized and his cargo of beaver pelts was confiscated.

Captain Bull was ruined, but not for long. He travelled to Boston, where he attempted to reacquire his property through the colonial court system. When legal means failed, the now enraged sea captain recruited a crew of shady characters from the Boston waterfront to help him recoup his losses by illegal means. Ironically, Bull soon decided to target British traders rather than French ship owners, probably because the former were wealthier.

Dixey Bull's piratical reputation stems from his attack on the Pemaquid settlement late in the summer of 1632. Bull boldly sailed his fleet of three ships into Pemaquid Harbor and sacked the town, including Shurte's fortified warehouse. The loot he seized was estimated to have been worth $2,500. Although he became famous along the Maine coast as a "dreaded pirate," there is also the hint that Bull may have been a seventeenth-century down east Robin Hood. He is reported to have adopted a series of measures to restrain his men against excessive drinking and despoliation. (How he did this we would like to know.)

A contemporary report of Bull's Pemaquid "visit" was noted by John Winter from the Richmond Island fishing station: "The last yeare [sic] was one that was when a trader for bever [sic] that is now turned pirate, and has done much spoyle [sic] here. His name is Bull. He tooke [sic] away from the plantation at Pemaquid as much goods and provisions as is valued to be worth five hundred pounds."

The authorities in Boston sent out a fleet of ships to find Captain Bull, but by 1633, as one historian wrote, "He sailed out of recorded history." Some stories say he joined the French; others say he was captured and taken to England, where he was hanged at Tyburn in London, although there are no records of this. There are also tales that he left buried treasure along the coast, perhaps on Damariscove Island and on Cushing Island in Casco Bay. Finally, there is the ballad that recounts Bull's death in a sword fight with a local Pemaquid fisherman named Daniel Curtis.

Over the centuries, "The Ballad of Dixey Bull" has grown to thirty verses, samplings of which follow:

Dixey Bull was a pirate bold.
He swept our coast in search of gold.
One hundred years have passed away,
since he cast anchor in Bristol Bay.

But Daniel Curtis, a fisherman,
Feared not the flag from which they ran,
But took his skiff; bent to his oar,
And rowed alone to Beaver's [islands] *shore.*

Curtis fought for cause that's right,
Dixey, because he liked to fight;
then down went Curtis upon his knee
The pirate's crew gave three times three.

Like a flash at him Dan went
And through his breast his sword he sent,
The blood gushed out warm, bright and red
The pirate staggered and fell dead.

Pirates, your flag and anchor pull,
for Curtis killed your Dixey Bull.
That's how Curtis won the day
And killed his man in Bristol Bay.

Treachery at Fort William Henry

Fort William Henry was the strongest fort on the East Coast in 1696. That year, however, it was the scene of a cowardly and treacherous act of betrayal by Pasco Chubb, the British officer commanding the newly built bastion.

Fort William Henry was located on the point of land that guarded the entrance to the Pemaquid River. The fort's strategic value was that it served as a buffer against French settlements located farther down east. As previously noted, Pemaquid was the first fortified English settlement on the coast of Maine.

In 1692, when King William's War (1689–1697) intensified in New England, Governor William Phips of Massachusetts spent £20,000, two-thirds of the annual budget of the Massachusetts Colony, to construct a large

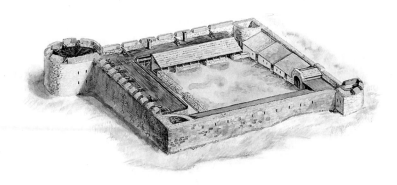

Fort William Henry was built in 1692 and fell in 1696. This is an artist's conception of the fort drawn by Alison Carver for the exhibit "Guns, Politics, and Furs." *Courtesy of Dr. Neill De Paoli.*

stone fort to protect the northern frontier of the colony from attacks by the French and Native American tribes. Workers used two thousand cartloads of stone to build a formidable fort—one with ten- to twenty-two-foot walls and a twenty-nine-foot-high watchtower. With a garrison of sixty soldiers and twenty cannons, Fort William Henry appeared to be impregnable.

Governor Phips wrote proudly, indeed arrogantly, of his accomplishment:

> *I have caused a large stone fort to be built at Pemaquid, and have kept a force ready to attack the Indians whenever they appear on our frontiers. The fort is strong enough to resist all the Indians in America and has so much discouraged their arms and sagamores [chiefs] to beg for an everlasting peace. I went to Pemaquid accordingly and concluded the articles of peace.*

William Phips was no longer governor when a force of Etchemins, Abenaki and French threatened Fort William Henry. The Indians were unhappy with the way they were being treated by the English settlers, and relations had deteriorated for a number of reasons. As the English population grew, they exerted greater pressure on the traditional hunting and fishing grounds of the Native American population. In their book *The Forts of Pemaquid*, Robert Bradley, Helen Camp and Neill De Paoli wrote, "Some traders in their urge to acquire prized beaver pelts, were less than scrupulous in their business practices. Native hunting practices changed rapidly and Indians faced potential starvation."

Under French leadership, the Etchemins and Abenaki surrounded the fort in 1696. According to one report, a sagamore's (chief's) son appeared with a flag of truce, apparently to request an exchange of prisoners, although other sources say they were demanding rum.

A French priest, Father Thury, warned of a trap, but two Abenaki chiefs, Edgeremet and Abinquid, went out to parlay with Captain Chubb. Then, in the words of nineteenth-century historian Samuel Drake, "Chubb took his advantage to lay violent hands on them." The English raised their guns and shot Edgeremet and his two sons, though Abinquid managed to escape. Needless to say, the Indians were enraged by Chubb's action and, together with the French, besieged the fort.

The supposedly impregnable Fort William Henry slowly crumbled under their persistent attack. The mortar used to hold the stones together was of poor quality, and the walls began to disintegrate. Moreover, the fort's water supply was outside the fort, which further limited the garrison's ability to withstand a siege.

Chubb's next traitorous action came a few weeks later, when he surrendered Fort William Henry, having been promised safe conduct by the French and Indian attackers. The garrison was left to fend for itself. One report says the defenders were massacred. The Indians destroyed the fort after finding another of their chiefs who was being held captive, "half dead in irons."

Pasco Chubb fled to Boston, where the angry governor interrogated him. He was stripped of his captain's commission and thrown in jail for surrendering Fort William Henry. When he was eventually released, the chastened Chubb crept away to live in seclusion with his wife in nearby Andover.

Two years later, the traitorous former Captain Chubb and his family were discovered living in Andover. On February 22, 1698, a group of thirty Abenaki attacked and killed the entire family to avenge the death of Chief Edgeremet and his sons. The villainous Chubb was reportedly shot several times through the head to make sure he was dead.

Other Pemaquid Forts

Fort William Henry was actually the third of four forts to be built on the Pemaquid peninsula. Prior to Fort William Henry, Fort Charles was built during King Philip's War (1676–1677) to protect the remaining English settlers in the region. The area had been left defenseless following a raid by French and Indians in 1676. At that point, New York's Governor Andros decided to bring New England under the jurisdiction of his colony. Accordingly, he sent an expedition to build Fort Charles, which was completed in 1677.

At full strength Fort Charles had a garrison of 150 men and was enclosed in a wooden stockade, a formidable stronghold for a frontier settlement. In 1688, however, the political situation in England changed when King James II was overthrown in the Glorious Revolution. As a protégé of King James, the unpopular Governor Andros was soon arrested and deported. At Fort Charles the result was that most of the soldiers immediately deserted. When King William's War broke out, only a handful of soldiers remained to protect the local population.

In August 1689, Edward Randolph, one of Andros's subordinates, wrote from where he had been jailed in Boston: "The Indians have overrun two hundred miles of coast. They have taken the town and fort of Pemaquid." The Pemaquid settlement was cut off from the fort so that only a few women and children were able to find refuge.

The French missionary who accompanied the Indians reported on the savagery of the assault of the town: "They put to flight all those in their nightclothes and threw themselves in their manner on the houses breaking down doors, taking and killing all whom they found inside."

At Fort Charles, Lieutenant James Weems wrote, "I was attacked by a great body of French and Indians, and having lost all but eight of my men I was obliged to capitulate." Fort Charles was destroyed, and the English abandoned mid-coast Maine for the next three years until the decision was made to build Fort William Henry.

Following the previously discussed destruction of Fort William Henry in 1696, the Massachusetts General Court, which was the colonial legislature of the Massachusetts Bay Colony, stubbornly refused to re-fortify the Pemaquid area. The reasons it cited included the expense and "strategic irrelevance of the location," i.e., there were no other English settlements in the area. As Robert Bradley reminds us in *The Forts of Maine, 1607–1945*, "Unfortunately, the Massachusetts Assembly (General Court) could not forget the expenditure of $20,000 wasted on Fort William Henry."

In the early eighteenth century, a new Massachusetts governor, Joseph Dudley, tried several times to persuade the General Court to allow settlers to move into the now deserted Pemaquid area. He wrote, "If there might be an establishment of an English colony at Pemaquid it would hold 1,000 families and would defeat the French and Indians hopes of resettling in those parts." Governor Dudley's efforts, however, were unsuccessful.

Further attempts to resettle the area failed, including the promise of building a bombproof barracks. It was not until thirty-three years later (August 15, 1730) that a letter appeared in the *Boston Gazette* written by the surveyor of His Majesty's Woods in America, Colonel David Dunbar. In his letter, Dunbar, whom Robert Bradley describes as "an able and aggressive administrator," advertised for settlers to repopulate Pemaquid and other former villages in the area. Dunbar soon found, however, that he was acting against the wishes of the Massachusetts General Court, which challenged his authority.

Undeterred, Dunbar imported Scots-Irish immigrants from Boston and established a settlement at Pemaquid late in 1730. In *The Forts of Pemaquid*, authors Bradley, Camp and De Paoli described what happened next:

> *His* [Dunbar's] *abrasive personality aside, such an action was bound to anger Massachusetts and the descendants of 17th century Pemaquid settlers, who saw their titles to the land being usurped. Dunbar lost his legal battle with Massachusetts and was forced to abandon the enterprise. This*

A replica of the stone bastion of Fort William Henry was built in 1908. *Author's collection.*

resolution of the dispute was devastating to many of Dunbar's immigrants, and it was the death-knell of the re-born village.

Ironically, the next year the unpredictable Massachusetts General Court decided to build a new fort, something it had resisted for the past thirty-six years. In 1732, Fort Frederick, named for the Prince of Wales, was built on the ruins of Fort William Henry. For the next twenty-seven years, the size of the garrison varied, depending on the state of relations with the French and local Indian tribes.

Finally, in 1759, the General Court decided that the site had become irrelevant due to the construction of Fort Pownal, at the head of Penobscot Bay, and the winding down of the French and Indian War. The remaining garrison was removed in 1761, and the fort fell into decay. On May 24, 1775, at the start of the American Revolution, local residents voted to "pull down Pemmaquide Foart" (Fort Frederick) to avoid its occupation by British troops.

Today, the tower of Fort William Henry, the partially rebuilt walls and the nearby Fort House are open to the public. Historian Neill De Paoli's impressive exhibition, "Guns, Politics, and Furs," can be seen on the second floor of the fort's stone tower, which also provides a magnificent view of John's Bay. The fort is open daily from Memorial Day through Labor Day.

CHAPTER 5

THE PENOBSCOT FORTS

The Contrarian of Fort St. George

Many years ago there was a saying: "As contrary as Old Johnny."
—Jim Skoglund, St. George Historical Society

On June 22, 1814, the seventy-four-gun British ship of the line *Bulwark* appeared off the mouth of the St. George River. The War of 1812 had been raging for two years, and HMS *Bulwark* was cruising the Maine coast, burning American merchant ships and plundering towns, including Castine and Belfast. Fearing his ship might run aground, *Bulwark*'s captain dispatched two barges to row eight miles up the St. George River with the expectation of adding the prosperous town of Thomaston to his list of conquests. Halfway up the river, they encountered Fort St. George, which is where our story begins.

Jim Skoglund is a retired history teacher and former state legislator who lives in the town of St. George. He is a member of the St. George Historical Society and is considered an authority on the history of Fort St. George. There are a number of restored forts along the Maine coast, although Jim reminded me that Fort St. George is one of the few to remain untouched.

From a plaque near the highway (Route 131) it is a quarter-mile walk through the woods to the fort, which is owned by the state and maintained by the Town of St. George. The fort sits on a pretty, grassy knoll overlooking the St. George River across from the town of Cushing. The two-acre site is

Aerial photograph of Fort Point and the ruins of Fort St. George, taken in 2007. *Courtesy of Christopher Leavitt.*

well off the beaten track and has been kept in its natural state. (The town mows the area just once a year.) As can be seen in the aerial photograph above, there are no places for a boat landing or access for cars.

From 1719 to 1759, the St. George River, also known locally as the George River, had marked the northeasternmost extension of English settlements in colonial New England. Steven Sullivan describes the situation in his 1987 master's thesis, "Settlement Expansion of the Northeast Coastal Frontier of New England":

> *A number of dynamic factors and circumstances interacted in the process of establishing and maintaining the St. George's frontier. This complex interacting of forces kept the frontier from either advancing or retracting during the forty-year period shaping the course and direction of English settlement expansion in central and eastern Maine.*

Fort St. George should not be confused with another, older, Fort St. George, located four miles upriver in the present-day town of Thomaston.

(Indeed, Maine seems to be filled with Forts George or St. George. There was also a Fort St. George in Phippsburg constructed in 1607 by the Popham Colony and a Fort George that was built by the British at Castine in 1779.) Fort St. George in Thomaston was built about 1720 and consisted of two blockhouses surrounded by palisades. For the next forty years, this fort sustained intermittent attacks by hostile Indians until the end of the French and Indian War in 1763.

With the end of the French and Indian War, the boundary of the frontier advanced farther down east to the Penobscot River. Fort Pownal, named after Massachusetts governor Thomas Pownal, was built and served as a buffer for the St. George River/Thomaston settlement.

When Henry Knox moved to Thomaston in 1794, he built his mansion, Montpelier, near the site of the original fort. In 1868, when Nathaniel Hawthorne visited Thomaston, he wrote of a visit he made to Montpelier, then in disrepair. Near the mansion and an old burial ground, Hawthorne recorded he saw an "ancient fort" that had been erected for defense against the French and Indians. This was Fort St. George, though no traces of it remain today.

By the end of the American Revolution, the children and grandchildren of the early settlers on the Cushing/Thomaston side of the St. George River had crossed the old frontier boundary line and claimed lots on the eastern side of the river. One of the settlers to "take up" a lot was Joseph Robinson, who brought in developers from Massachusetts to do the surveying.

Constructing the Fort

As relations between Great Britain and the United States deteriorated in the early nineteenth century, President Thomas Jefferson issued the Embargo Act in 1807, hoping to head off war between the two countries. The purpose of the act was to reduce tensions with Great Britain by prohibiting American ships from leaving their ports. The result, however, was a commercial disaster for the maritime community. Although Jefferson repealed the hated act before he left office, the Embargo hit New England shippers particularly hard. Desperately in need of sailors for their life and death struggle with the French, Great Britain continued to stop American ships and impress sailors, especially those suspected of being British citizens.

Relations with England continued to worsen, and in 1808, Congress authorized Secretary of War Henry Dearborn to construct a series of

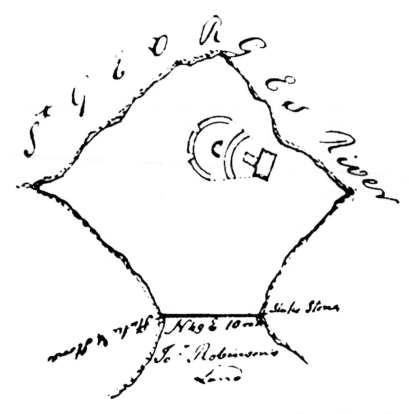

Sketch of Fort Point and Fort St. George drawn in 1808 by John Gleason. *Courtesy of St. George Historical Society.*

forts along the Maine coast. The fear was that Great Britain might seek retribution for the Embargo Act. An army engineer, Major Moses Porter, was directed to oversee the construction of four forts, one of which was Fort St. George. Fort Edgecomb on Davis Island, across the river from Wiscasset, was another.

By the early 1800s, the town of Thomaston had become a thriving maritime hub, which made it an attractive target for the British. With the old Fort St. George at Thomaston in ruins, it was critical that another fortification be built to protect the area. By the time President James Madison declared war on Great Britain in June 1812, Fort St. George was completed and garrisoned. Although only a modest, semicircular earthwork, it was hoped that the fort's location on a rocky promontory facing the river would discourage the British warships from venturing up the St. George River.

In his book *Maine Forts*, Henry Dunnack described Fort St. George's excellent location:

> *Standing on Fort Point, which rises gradually in a smooth field to the river, there is a view of great beauty. Looking up the river one sees Thomaston and the Camden Mountains. Looking down the river towards Muscongus Bay one sees the shores of Friendship and at night Monhegan Light, twenty-four miles away, can be plainly seen.*

The United States government purchased the land for Fort St. George from Joseph Robinson on September 22, 1808, as recorded in the deed records of Lincoln County. Hezekiah Prince, a local businessman and a first selectman of St. George, wrote in his diary that on June 28, 1809, Major Moses Porter arrived to oversee the construction of the fort. Prince, whose many talents included those of a skilled carpenter, was subcontracted, along with Captain Thomas Vose, to do the actual construction.

Prince's diary notes that he spent a week getting loads of sod and timber for the fort. According to Cyrus Eaton's book *A History of Thomaston, Rockland, and South Thomaston, Maine* (1865), Prince built a rampart in the form of a crescent toward the river, upon which were mounted two or three eighteen-pound guns. There is some question as to the exact number of cannons. John Locke records in *A History of Camden and Rockport* (1859) that there were four cannons, as well as two brass artillery pieces. Eaton adds, "Attached to this were the barracks, a small blockhouse and a magazine of brick. The enclosure was completed in the rear by a high board fence." Captain Thomas Vose, a veteran of the Revolutionary War, a resident of Thomaston and a business associate of the recently deceased Henry Knox, was put in charge of the fort until he died in 1810.

With the outbreak of war in 1812, Fort St. George was used as a recruiting center and drilling ground for local militia under the command of a Sergeant Nute. After a few weeks of training, his men were dispatched to other locations. Disappointed by the disappearance of his command, Nute turned the fort over to Hezekiah Prince.

As previously mentioned, Prince was a local entrepreneur involved in various business enterprises; in 1814 he would become manager of the Thomaston Cotton and Woolen Manufacturing Company. With a now empty fort and preoccupied with his commercial activities, Prince placed Fort St. George in the hands of Ephraim Wiley, a fifty-eight-year-old veteran of the Revolution who had served at Valley Forge. Wiley was told to "stay there and keep things in order."

"Leave the Premises"

And so we return to the night of June 22, 1814. The story goes that the fort's solitary occupant, Sergeant Wiley, had already finished his supper when British marines came ashore. They spiked the cannons, and in a loud voice the officer in charge demanded the fort surrender "in the name of King George." Hearing no response, the officer ordered a musket to be fired at the barracks door. The ball passed through the door's upper panel and grazed the shoulder of Sergeant Wiley, now lying in his bunk. At this point, Wiley opened the door and ordered the British to leave the premises.

Ephraim Wiley's headstone. Wiley was known as "the cantankerous caretaker of Fort George." *Author's collection.*

In a scene reminiscent of the Keystone Kops, the British officer, taken aback by this response, asked to see the commander of the fort. The defiant Wiley snapped, "I am the commander. This is Hezekiah Prince's fort and he has left me in charge." Wiley was then ordered to step forward and surrender the flag. Far from being intimidated, the angry Wiley replied, "I told you once, this is Squire Prince's fort, and if you want any flag, you must go to Squire Prince."

The British left, having had their fill of the cantankerous old caretaker. Their plans to blow up the fort were also thwarted, as they were unable to find sufficient gunpowder. The meager supply they did find they scattered to the winds. The irate Ephraim Wiley was left unharmed except for a slight wound in his shoulder. Thus, Fort St. George's single military "engagement" came to an end.

The British marines crossed the river and took out their frustrations on the residents of Cushing by setting fire to two vessels and, according to Cyrus Eaton, "engaging in a brisk fire" with the residents of the town. A young man, Christopher Curtis, was discovered trying to escape on one of the boats. With a pistol to his head, he was told to pilot the British barges upstream, where the town of Thomaston lay undefended.

There should be a monument in Thomaston to the quick-thinking Curtis. Initially, the youth dared not refuse the order, and he guided the marines up the St. George River toward Thomaston. When they approached to within five hundred yards of the town, Curtis told his captors that they still had "a very long way to go." In the darkness and fog the British believed the cagy young man. The expedition was abandoned and Thomaston was saved.

If there is a postscript to the brief "capture" of Fort St. George, it occurred two months later when an alarm, which turned out to be false, was raised that the British were again coming up the river. Humiliated by having been unable to fire a shot, Ephraim Wiley planned his defense. He had already unspiked the guns and recovered enough of the scattered powder to load three cannons. Apparently, Wiley was ready to take on the British single-handedly if they had the nerve to return. Fortunately for the old fellow they never did, but he fired his cannons off anyway to help spread the alarm.

Old Johnny's Claim

On March 27, 1874, John Robinson, nephew of the fort's original owner, Joseph Robinson, filed a pension request with J.E. Wheelock at the local pension office. After the War of 1812, the fort fell into ruins, although "Contrary Johnny" continued to farm the land and mow the fields. "Old Johnny," as he was also called, was eighty-seven years old, and he was requesting a pension for services he performed during the War of 1812. Robinson's pension claim #13847 can be seen today. An approximation of his service record follows.

Sometime in July 1814, while he was at home hoeing corn, Robinson received orders from the commander of the town militia, Captain Kinney, in whose company he was an orderly sergeant. Robinson thinks it may have been July 21. Whatever the exact date, the fear of an enemy attack was still very much on people's minds.

Robinson was directed to join his company in front of Squire Prince's store in St. George. He was then told to take a guard of four men to Cobb's Narrows, which is downstream from Fort St. George. He remained there until September 1, when Kinney ordered him to "march his men to Camden."

Fort historian Jim Skoglund told me "this was a big deal." Robinson, who appears to have been designated as quartermaster for the company, was ordered to bring provisions for fifty men, who then walked the fifteen miles to Camden. The militia stayed at Camden for a week before marching back.

Robinson was ordered to return to Cobb's Narrows, where he stood guard duty for another forty days. Robinson claims that during this time he frequently visited Fort St. George. Skoglund told me the whole event was similar to closing the barn door after the horse got out, since the British had already come and gone.

Back to the pension claim: Robinson got Jane Spear from Rockland, age seventy-seven and widow of William Spear, to support his claim. Her husband had been mustered into service at the St. George fort in August 1814, and she recalled him serving in the same company as Ensign John Robinson. She completed her testimony by saying that Robinson was "still alive and living in St. George." Skoglund said it was "remarkable that in 1874 she could remember who her husband served under sixty years previously."

The record also shows that Aaron Mossman, from Union, Maine, remembered that John Robinson commanded the guard station at Fort St. George for three months, "commencing the last of July." Mossman said that his knowledge was "personal" since he served under Robinson.

Philip Sukeforth lived in Appleton, Maine (north of Union), and called himself "a survivor of the War of 1812." Sukeforth said that he served under John Robinson while at Fort St. George for a period of three months, starting at the end of July 1814. Sukeforth reported, "Robinson commanded the guard station where he held the rank of ensign."

Sukeforth goes on to say that "he remembered Robinson well since he got into an altercation with him and promised to 'pound

Headstone of "Contrary Johnny." He finally received a pension in 1871 for serving in the War of 1812. *Author's collection.*

him,' but Sukeforth left the aforesaid, unbeknownst to him [Robinson] before this could take place." Jim Skoglund thinks this was a rare testimony in view of the fact that Sukeforth supported Robinson's claim for a pension.

Presumably, Robinson received his pension. Congress had passed a new pension act in 1871, which granted pensions to surviving soldiers and sailors who had served sixty days in the War of 1812 and had been honorably discharged.

Present Day

The earthworks of Fort St. George remain, and parts of the foundation for the barracks and blockhouse can be seen today. For many years, there were ringbolts for the cannons that could be moved around the ramparts. In the 1870s, deep holes were dug behind the fort where it was thought pirates had buried treasure, though none has ever been found.

As a boy, Jim Skoglund remembers meeting the great-grandson of Old Johnny Robinson. As Jim told me, "I feel a connection with the old guy." Jim is pleased that he knows someone who remembers the people who served here. "It makes everything seem a lot closer." As has already been noted, John Robinson was an ornery character, which gave rise to the local saying, "He was as contrary as Old Johnny."

On March 4, 1923, the 2.63 acres on which Fort St. George stood were appraised and sold to the State of Maine for the bargain price of $22.50. Governor Percival Baxter sent a check for that amount to Secretary of War Robert C. Davis. The site of Fort St. George became a state memorial on December 4, 1929.

THE DEFENDER OF FORT GEORGE

British sea captain Henry Mowat was one of the more controversial characters in the American Revolution. Some consider him a "war criminal" for ordering the destruction of Falmouth (Portland) in October 1775. George Washington would later write of his action, "I know not how to detest it." While American Revolutionaries saw him as a destroyer, Loyalists considered him the heroic defender/savior of Fort George and the town of Castine.

Jim Stone is a retired Wesleyan history professor and a member of the Castine Historical Society. Professor Stone points out that the Mowat story is complex. To begin with, the Scottish-born captain was following orders from his superior, Admiral Samuel Graves, commander of British naval forces in North America, who instructed him to destroy as many seacoast towns as he could.

The Background

Following the Battles of Lexington and Concord, rebel forces captured Mowat near Falmouth on May 7, 1775. He was released, having given his word of honor to his captors that he would return the next morning. Mowat, however, broke his parole and immediately took his ship, *Canceaux*, to Boston. The next fall, Admiral Graves ordered Mowat to "proceed along the coast and lay waste, burn and destroy such seaport towns as are accessible to His Majesty's ships...to make the most vigorous efforts to burn the towns and destroy the shipping." Graves added that it was up to Mowat to go wherever he wanted.

In fact, Mowat disobeyed his orders. He arrived in Casco Bay on the evening of October 16, 1775, in command of three small warships. The next day, he sent a barge to Falmouth with a letter to the town fathers stating that he had orders to fire on the town immediately. Mowat added that he would "deviate from his orders" and give the townspeople two hours to evacuate the next morning. Captain Mowat emphasized that his orders could not be changed and that he was already risking the loss of his commission by giving the town a warning. It turned out to be a decision that would haunt him for the rest of his career.

Mowat actually waited an additional half hour before beginning to bombard Falmouth at 9:40 a.m. on October 18. Then at three o'clock he sent landing parties ashore to set fire to buildings not demolished by gunfire. It is estimated that over four hundred structures were destroyed, leaving most of the town homeless. Professor Jim Stone writes, "Mowat's name would go down in infamy." Portland journalist C.E. Banks, 130 years later, compared Mowat to Nero: "His unparalleled barbarity was exploited abroad and his name finally consigned to that limbo of hopeless condemnation where he will be remembered by future generations as a fiend and not as a man."

Although Mowat would prove to be the indispensible man in the defense of Castine in 1779, his name has been somewhat tarnished by the notoriety he received as the destroyer of Falmouth, as well as for having broken his parole. Nor was his name respected as a British naval officer—from a military perspective he was seen as disobeying Graves's orders by warning the town of his imminent attack. The result was that for the rest of his career Mowat was denied promotion due him as a captain in the Royal Navy, and it was none other than King George III who ultimately blackballed Mowat's career. Following his defense of Fort George in 1779, he would remain on station in Penobscot Bay as commander of a small naval contingent until the end of the war, in what Jim Stone calls "professional obscurity."

Colonial Castine Was Called Pentagoet

We will return to a further evaluation of Mowat's career in due course, but first it is important to examine the early history of Castine. The location first appears on a 1612 chart drawn by the great French explorer Samuel de Champlain, who called it the Pentagoet Peninsula. Beginning in 1613, French traders and missionaries set forth on their expeditions from a small, fortified trading post named Fort Pentagoet.

It is hard to believe the charming resort town one sees today was once so important militarily. Located at the head of Penobscot Bay and guarding the entrance to the Penobscot River, Pentagoet's strategic position was what appealed to the French in the early seventeenth century. In fact, Pentagoet (it was incorporated as the town of Castine in 1796) was the first permanent settlement to be built in New England.

Because of its location next to a deep, sheltered harbor and its access to timber and wealth from the fur trade, Professor Stone told me that the Pentagoet Peninsula became the single most contested piece of property (one and a half square miles) in the entire world. For the next two hundred years, the town would change hands seventeen times. "It was originally a pawn in the diplomatic game and it very quickly became a rook," said Stone.

For the next forty years, control of the Pentagoet Peninsula shifted back and forth between England and France. Baron Jean de Saint Castin arrived in 1667, armed with a land grant from France's King Louis XIV. Castin, for whom the town is named, married the daughter of a local Abenaki chief who bore him ten children. Meanwhile, Fort Pentagoet was becoming a prosperous trading post.

The Dutch briefly captured Fort Pentagoet in 1674 as a part of the larger struggle between France and the Netherlands. A Dutch privateer sailed into the harbor and, finding only a few French defenders, easily took the town. Dutch control of the area was never fully established, however, and two years later Baron Castin retook the settlement. For the remainder of the seventeenth century, France and England continued to vie for control. Despite frequent conflicts with the English, Castin returned to France in 1703 a wealthy man, leaving his son Joseph in charge.

French control of the Pentagoet Peninsula waned during the first half of the eighteenth century until control was permanently transferred to Great Britain at the end of the French and Indian War. The lands were opened to settlement, and colonists, mostly from Massachusetts, began to move into the unoccupied lands along the Maine coast. Although the fur trade had dried up, the fishing and timber industries prospered. The British government was particularly interested in timber the area could provide for its growing navy.

Mowat and the Penobscot Expedition

Twenty-four-year-old Henry Mowat first came to North America in 1758. There he remained, serving in various capacities until the end of his life in 1798. Mowat came from good maritime stock. He was the son of Captain Patrick Mowat, who commanded a ship on Captain James Cook's first global voyage in the 1760s.

At the conclusion of the French and Indian War in 1763, young Mowat was ordered to guard a survey expedition for the official cartographer of King George III. For the next twelve years, Mowat guided royal survey ships up and down the Atlantic coast from Newfoundland to the Virginias. By the start of the American Revolution, Professor Stone writes, "His [Mowat's] many years on the survey of the coast to the eastward of Boston and his knowledge of all the harbors, bays and creeks and shoals resulted in his being the most knowledgeable and experienced naval commander in British North America, bar none."

Mowat's destruction of Falmouth (Portland) in October 1775 has already been discussed. For the next four years, the notorious captain prowled the New England coast in the sloop of war *Albany*, preying on American shipping. He took prizes and protected the sea lanes between New York and Halifax after the British evacuation of Boston.

In 1778, Sir Henry Clinton, commander of British forces in North America, was told by Lord Germain, minister for the colonies, of a plan to establish a new British colony. The province would be named New Ireland,

and it would encompass lands running from the Kennebec River to the St. Lawrence River. The new colony would serve as a buffer between Nova Scotia and New England and act as a haven for American Loyalists. (The British lord chief justice would eventually reject the idea because it violated an English law by taking land from one Crown colony and giving it to another.)

Following up on Lord Germain's plan, in February 1779, Henry Clinton sent secret orders to British general Francis McLean stationed in Halifax: "To establish a settlement and erect a fort on the Penobscot River…Prepare materials for a respectable work capable of containing three or four hundred and consult Captain Mowat on this subject." This pleased McLean, who wrote that he would "derive much assistance from Mowat's abilities and knowledge of that coast."

Mowat arrived in Halifax on schedule to convoy the expedition. In June 1779, a fleet of three warships and four armed transports carrying 650 Scots soldiers set sail from Halifax. Henry Mowat commanded the fleet in his flagship, *Albany*. The expedition's purpose was to establish an outpost on the western boundary of New Ireland that would become a place of safety for Loyalists. The fort would be located in present-day Castine and named Fort George in honor of King George III.

Led by General McLean, the soldiers disembarked on June 17, 1779, and took control of what at that point was a mere hamlet surrounded by several sawmills. A site for the fort was staked out on a commanding rise two hundred feet above sea level, and work was begun.

Professor Stone points out that an added advantage was that most of the officers and men were Scots, who were well known for their engineering skills. Located above the village, Fort George was sited so that its guns had an open field of fire in any direction. In the words of George Washington, Fort George became "the most regularly constructed and best finished of any fortress in North America."

Building Fort George turned out to be a race against time. Word of the British occupation of the Pentagoet Peninsula reached Massachusetts within two weeks. The Massachusetts Board of War realized the strategic importance of the move and ordered that an expeditionary force be dispatched immediately to dislodge the invaders. The British occupation of the peninsula was too important to go uncontested, providing as it did strategic supplies including valuable sources of timber for the American war effort.

Space does not permit a detailed discussion of the ruinous Penobscot Expedition that followed. Suffice it to say that historians consider it one of the worst naval operations in American history. From the start, the mission

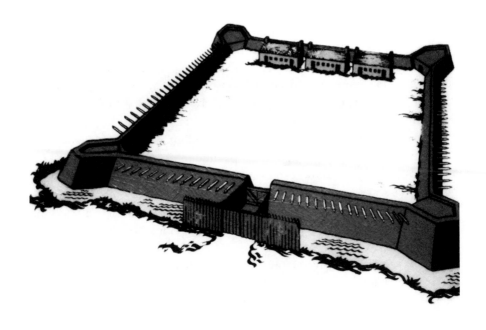

An illustration of what Fort George might have looked like is shown on a plaque outside the present-day fort. *Author's collection.*

was plagued with problems. Dudley Saltonstall, a cranky commodore from Connecticut, was put in command of a fleet of eighteen warships and twenty-four transports. General Solomon Lovell was in charge of one thousand poorly trained and ill-equipped militiamen and four hundred marines. A familiar name, Paul Revere, was given command of artillery. In their anxiety, the Massachusetts War Board gave Saltonstall and Lovell a mere week to get organized. The largest amphibious expedition of the American Revolution arrived off of Castine on July 25, 1779.

Meanwhile, British soldiers and American Loyalists living on the Pentagoet Peninsula, also known by its Indian name of Bagaduce, were feverishly clearing the land and constructing a quadrangular-shaped fortress made mostly of earth and sod. The earthworks acted as a cushion. They actually made Fort George less vulnerable because, in a short siege, cannonballs fired from ships would simply sink into the soil and not send splinters from wood and stone flying in every direction.

Captain Mowat's knowledge of upper Penobscot Bay was proving to be invaluable. Indeed, his reputation had preceded him, as it soon was evident that Commodore Saltonstall was intimidated by the very mention of his name. (As already noted, Professor Stone considers Mowat "the most knowledgeable British commander on the continent.") Even though the advancing American fleet, with 344 mounted guns on eighteen warships, outgunned the British, Saltonstall delayed an attack for several days. As a result, the British were able to continue work on Fort George.

Mowat was later to write, "These three or four days of embarrassment on the part of the rebels gave our troops time to do something more to the fort, to carry up the most necessary stores, and to mount several more guns." During these early days of the siege, Mowat "seconded" sailors under his command to work on the construction of the fort and artillery battery on the peninsula, with one of the batteries being manned by gun crews from his sloops.

At the same time, Mowat saw the long-range strategic importance of successfully defending the peninsula: "A British loss at Penobscot would have been the equivalent to a second Burgoynade [John Burgoyne was the British general defeated at Saratoga in 1777] and an immense encouragement for the Americans who were tiring of the length of the war."

British fleet advancing up Penobscot Bay—the American fleet is shown fleeing in the distance. *Courtesy of National Maritime Museum, Greenwich, England.*

To defend the peninsula, Mowat came up with a clever tactic. He strung, or "sprung" (the nautical term), his three sloops across the entrance to Castine harbor. This would force American ships attacking the town to face a concentrated broadside. As mentioned, Mowat unloaded his guns on the off side of his ships and sent them ashore, providing another battery for the defense of the peninsula.

For the first two weeks of August, Fort George was under an intermittent siege as the American commanders Lovell and Saltonstall argued over how best to proceed. The two men disagreed about everything—who was in charge, when to attack and where to land. It was August 13 before the cautious and indecisive commanders agreed on a plan of action. Saltonstall finally responded to a direct order from his superiors in Boston, who had been informed of the delays by his exasperated junior officers.

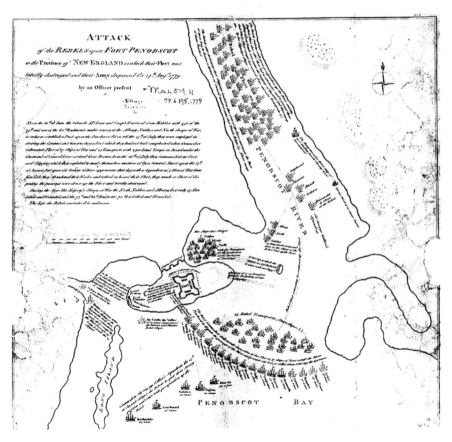

Chart showing relative positions of the two fleets and the American retreat up the Penobscot River. *Courtesy of Castine Historical Society.*

As the combined land-sea American operation was about to be launched, a British relief fleet under Commodore George Collier was spotted heading up Penobscot Bay toward the besieged fort. Collier's flagship, the sixty-four-gun *Raisonable*, was clearly superior to any of the American warships. Although the Americans had more warships and more guns, they quickly realized they were no match for the better-trained British crews.

The battle was over before it had hardly begun. After exchanging a few rounds with the British, the American warships turned tail and fled up the Penobscot River, followed by twenty-four ungainly transports crammed with troops. The next day, the river was filled with scuttled ships and the banks were littered with the remains of vessels that had been run ashore and burned. It took a month for the bedraggled American militiamen to make their way back through the wilderness to Massachusetts. After he returned from the ordeal, American general Lovell was heard to say, "Why couldn't we have fought the way the British did?"

Professor Stone describes Henry Mowat as a "heroic figure" throughout the whole affair. As commander of the three little sloops that held off the attack on the fort until the British fleet showed up, he provided a crucial delaying action. In his report to the admiralty, General McLean mentioned the invaluable actions of the three sloops of war during the siege and especially the measures taken by Mowat.

On the other side of the coin, Commodore Collier's dispatches to London failed to make any mention of the service of Mowat's little fleet. In short, Mowat's superiors agreed with an American historian's version of the affair that "Mowat's contribution to the British victory at Penobscot was negligible." Clearly, Mowat's noncompliance of his orders at Falmouth back in 1775 had not been forgotten by the British high command.

"Have You Paid Your Taxes?"

After the disorderly American retreat in August 1779, a British force remained at Fort George until well after the Revolutionary War. With no enemy challenges, there was nothing to do but work on the fort, which the Scottish soldiers did with apparent enthusiasm. More cannons were mounted, the walls were strengthened and a brick magazine, barracks and officers' quarters were built.

When the British finally abandoned the fort in January 1784, an American representative had not yet arrived to accept their surrender. The British

Remains of earthworks at present-day Fort George. *Author's collection.*

therefore burned Fort George before they left in "an indignant huff," according to Robert Bradley in his book *The Forts of Maine, 1607–1945.* Jim Stone adds one could therefore argue that Fort George was never legally or politically surrendered to the Americans. When he told the story recently to a group of tourists from the British Isles, a Scotsman asked him, "Well, have you paid your taxes?" The point being that the area was still technically British territory.

When the War of 1812 broke out, British troops returned to Penobscot Bay and recaptured Fort George from the Americans. Jim Stone said that the British fortified the whole peninsula much more strongly in 1812 than they did during the Revolution, although the fort would see no action. The walls were "redressed" (repaired), the garrison strengthened and additional cannons mounted. The now venerable bastion remained under British control until April 1815, when news of the peace treaty ending the war reached Castine. At that point, the British once again burned the wooden buildings and left.

Captain Henry Mowat, the Defender of Fort George, served in the British navy for forty-four years, most of which were spent patrolling and mapping the North American coast. At the end of the Revolution, however, Mowat's career languished. With the threat of Napoleonic France, the United States became a geo-political backwater for Great Britain.

Mowat was onboard his ship *Assistance*, five miles from Cape Henry, Virginia, when he was stricken with apoplexy on April 14, 1798. He died shortly afterward at the age of sixty-four. Henry Mowat was buried in St. John's churchyard in Hampton, Virginia, where his broken, moss-covered tombstone can still be seen lying on the ground.

THE SERGEANT AT FORT KNOX

For thirteen years, a Hungarian immigrant was the sole occupant of Fort Knox, the largest and most elaborate fort ever constructed in the state of Maine. Sergeant Leopold Hegyi was born in Budapest, Hungary, in 1829. It is thought that he came to the United States following the Revolutions of 1848–1849 that wracked much of Europe.

Hegyi probably first lived in the large Hungarian community in Brooklyn, New York. He joined the United States Army on October 4, 1867, and as a skilled horseman, he was assigned to the Mounted Service in St. Louis, Missouri. Before he moved to Maine, Hegyi spent the next twenty years of his life in Missouri training young men to become members of the United States Cavalry.

The Aroostook War

Well before Leopold Hegyi arrived in the United States, tensions were increasing on the border between Maine and the British Canadian province of New Brunswick. The result was the Aroostook War (1838–1839)—the only time an individual state has ever declared war on a foreign country. This occurred when Maine's legislature authorized Governor John Fairfield to declare war on England.

Following the American Revolution and the War of 1812, the border between Canada and the United States had never been clearly defined. When Maine became a state in 1820, the lack of a definite boundary with Canada was of major concern for the new state. One of the important issues was the question of timber rights. Lumbermen from both New Brunswick and Maine accused each other of logging on their lands. Militia was mobilized, and tensions ran high on both sides of the border until the United States and British governments stepped in.

The Aroostook War (also discussed in the "Boundary Forts" chapter) was named for a river on the United States–Canada border. The extent of the fighting, however, was a barroom scrap in the border town of Houlton, where both British and American troops were drinking. When someone offered a toast to Maine's success, a brawl broke out that resulted in black eyes and bloody noses. The border disagreement was not resolved until 1842, when Daniel Webster and Alexander Baring, Lord Ashburton, negotiated the Webster-Ashburton Treaty, ending the dispute.

Looking back on over a century of harmonious relations between the United States and Great Britain, it is hard to realize the tensions that once existed between the two countries. The British had attacked Maine several times during the American Revolution and more frequently during the War of 1812. People living along the Penobscot River and in Bangor, a booming lumber and shipbuilding area, felt increasingly vulnerable in the event that war again threatened.

Before the Webster-Ashburton Treaty was even concluded, Congress appropriated $25,000 to build "defensive works, barracks and other necessary buildings" in the Bangor area. A site was purchased, south of Bangor, on 125 acres of high ground across the Penobscot River from Bucksport. Enemy ships heading up the river would have to pass "the narrows" at Verona Island, where they would be exposed to shore batteries.

A Third System Fort

In the middle of the nineteenth century, the chief engineer of the United States Army was West Point graduate Joseph Totten. Until his death in 1864, Totten was directly involved with the design and construction of Third System forts, many of which, including Fort Sumter, came under fire during the Civil War.

Simply put, the Third System of seacoast defenses were multi-tiered structures with numerous guns concentrated behind high, thick masonry walls. Fort Knox, for example, had mounts for 130 guns. (First and Second System forts were less complex fortifications built earlier in the nineteenth century.)

The present-day Fort Knox parade ground. The towers of the Penobscot Narrows Bridge can be seen in the background. *Author's collection.*

Although Totten drew the plans for Fort Knox, army engineers Isaac Stevens and Thomas Casey supervised the actual construction. As noted in the "Portland Harbor Forts" chapter, Casey went on to an impressive career, eventually succeeding Totten as chief of army engineers. He took over the stalled construction of the Washington Monument in 1879, which was completed in 1885. It is interesting that the towers on the nearby Penobscot Narrows Bridge are modeled after Casey's completed Washington Monument.

Surveying and excavation of the Fort Knox site began in 1844, and construction proceeded, albeit intermittently, for the next twenty years. John Cayford wrote in *Fort Knox, Fortress in Maine*, "Beautiful granite from nearby Mount Waldo [five miles farther up the Penobscot River] began flowing into the work site. The greatest difficulty was in transporting the granite blocks."

Dick Dyer is a former Fort Knox ranger who runs a public relations firm in Winthrop, Maine. Dyer added, "When the huge granite blocks were winched out of the quarry, they were dragged by oxen to barges on the river. They were then floated down the river to docks in front of the fort. The docks are still there."

Although nearly $1 million was ultimately spent, Fort Knox was never quite finished. Dick Dyer described the enormous fort as "state-of-the-art for its day, although by 1900 it was an irrelevant citadel." A steam derrick was used to lift heavy objects from the barges to the docks. At the land entrance to the fort there are remains of tracks for an apparatus that carried materials down into the fort. Dyer described the engineering involved in some of the stonework as "amazing, especially the two hand-chiseled spiral stairways."

Fort Knox was never attacked. With granite walls twenty feet high and forty feet thick, it would have presented a formidable challenge. Once inside an outer wall, attackers would have had to cross a dry moat where they would be exposed to murderous rifle fire. On the Penobscot River side there were eight massive casements for heavy guns, all of which supported additional guns mounted above. The granite walls were so thick that ships' cannonballs would have been ineffective.

I asked the former ranger what he could tell me about the fort that wasn't in the guidebooks. Dyer told me that most people didn't realize that Fort Knox was built on a granite vein that runs under the Penobscot River to the town of Bucksport. During a Fourth of July celebration one year, all the fort's guns were fired simultaneously. The vibrations broke a great many

Fort Knox as seen from Bucksport, across the river, in the mid-nineteenth century. *Collections of Maine Historical Society, Courtesy of Maine Memory Network.*

windows in buildings across the river. "In fact, it happened twice," the old ranger admitted, "and it was rather embarrassing."

Then there is the story of a winch system that was connected to a huge chain running over to Verona Island. In the case of an attack, the chain could be raised up to stop advancing vessels and leave them vulnerable to fire. Dyer said this was never confirmed: "There were buildings on the shore, but none of them had a winch system. If one was there it wasn't in the architect's plans."

Did a tunnel run under the river to the other shore? Dyer speculates that the idea may have stemmed from the fact that a large drainage tunnel led down to the river's edge from the fort. He remembers when he was a ranger that boys would crawl into the culvert, thinking that it led to a secret passage across river. Frequently he had to go in and pull them out. "Given the engineering skills at the time," Dyer said, "it would have been quite a feat to cut a tunnel through the granite across the river."

Fort Knox was named in honor of Henry Knox, America's first secretary of war. Knox "retired" to Maine in 1794 and built a mansion (Montpelier) in Thomaston. He followed this up with ambitious operations in lumbering, shipbuilding, brickmaking and cattle raising. Knox also constructed the first canal with locks on the Georges River before his untimely death in 1806 at the age of fifty-seven.

The Civil War Era

Although Maine was never threatened by a Confederate invasion, there were several instances when citizens of the state were touched by the war. One such time was the 1863 raid on Portland Harbor by Charles Read, a notorious Southern privateer commander. Read was apprehended as he was trying to escape with a Federal revenue cutter he had hijacked. Then there was the plot in 1864 to rob a bank in Calais, Maine. It was discovered in the nick of time, and the robber, a Southern officer seeking to "fundraise" for the Confederacy, was caught and arrested.

Leon Seymour is executive director of the Friends of Fort Knox, and he told me the three reasons why Fort Knox was important during the Civil War. The first was that until the middle of the war, there was no guarantee that Great Britain would not enter on the side of the Confederacy. Seymour emphasized that had Great Britain joined the South, Fort Knox would have been, in modern parlance, "ground zero." British troops invading from Canada would have hit Fort Knox first.

The Rodman gun weighed twenty-five tons and could fire a 450-pound shot three miles. *Author's collection.*

A postcard of Rodman guns (Battery B), Fort Knox, circa 1930. The "hot shot oven" can be seen on the right. *Collections of Maine Bureau of Parks and Land, Courtesy of Maine Memory Network.*

Secondly, Fort Knox was very well armed. Of particular note were its twenty-four flank howitzer cannons, more than in any other American fort. Conceptually, those weapons sprang from a need to cover the blind approaches and moats surrounding the fortifications; in other words, they protected the flanks of the fort. The flank howitzer had no handles but was placed on traversing carriages, which greatly increased its field of fire.

Finally, Seymour told me that Fort Knox, armed with four huge Rodman guns, was the "nuclear missile silo" of its day. A dozen men were required to load the fifteen-inch muzzleloading monsters. Each gun weighed twenty-five tons and could throw a 450-pound shot three miles when elevated to twenty degrees. Designed especially for seacoast fortifications, the Rodman guns were the most powerful weapons in the war.

Fort Knox was nearing completion when the Civil War broke out, which caused further construction to be put on hold. Instead, the fort was used as a training base for artillery and infantry units from Maine. Records show that the complement of men training ranged from twenty to sixty troopers. At the end of the war, Fort Knox was again turned over to the corps of engineers, with orders to finish the job. Construction proceeded sporadically until it finally stopped for good in 1869, with the fort still unfinished.

An "Excellent" Professional Soldier

Let us return to the career of Sergeant Hegyi. He joined the army in 1867, where his talents as a trainer of cavalrymen and their horses were quickly noted. John Cayford tells us, "Leopold Hegyi accomplished a feat during his first five-year enlistment which few enlisted men have achieved. He was promoted to the rank of First Sergeant, one of the highest non-commissioned ranks in the U.S. Army. Throughout his long military career, Sergeant Leopold Hegyi was consistently given an excellent rating as a professional soldier."

After completing his initial five-year tour at the St. Louis Depot, Hegyi was transferred to the renowned Jefferson Barracks, also in Missouri. In the middle of the nineteenth century, the Jefferson Barracks was one of the largest military posts in the United States and one with a reputation for producing the best-drilled troops in the United States Army.

For the next fifteen years, the Hungarian-born sergeant trained young recruits at the Jefferson Barracks. Most of the troopers were headed for remote army outposts farther west, where their duties frequently included oversight of rebellious Indian tribes. Included among his trainees were troopers in General George Custer's ill-fated Seventh Cavalry.

When Leopold Hegyi reenlisted in 1887, he learned that he had been assigned to Fort Knox in Maine. En route to his new post he stopped in New York to visit friends—this is also where he married his wife, Louise. Little is known about when they first met, but what *is* known is that when Hegyi took up his duties at Fort Knox, his wife was not with him. We also know that twice a year, Louise Hegyi and her two small dogs took the train to visit her husband at Fort Knox. Her visits lasted two weeks, and she boarded at the home of Edwin Grindle in nearby Stockton Springs. Mildred, one of the Grindle children, remembers playing with the dogs, until one of them snapped at her. She also remembers the Old World charm of Leopold Hegyi.

John Cayford calls Ordnance Sergeant Leopold Hegyi "The Lone Guardian," referring to his thirteen-year stint at Fort Knox at the end of his life. Hegyi was a twenty-year veteran of the United States Army by the time he received his orders directing him to Fort Knox. He was designated "Fort Keeper," in charge of maintaining and guarding government property.

The fort had not undergone any repairs for almost twenty years, and John Cayford speculates, "It must have been disappointing to Sergeant Hegyi when he looked over the deserted garrison for the first time. It was in a state of disrepair; windows broken, roofs leaking, rooms piled high with debris." What did impress Hegyi, however, was the intricacy of the stonework in the fort. He compared it to the best in Europe.

John Cayford describes Sergeant Hegyi's daily routine:

You woke at daybreak, washed, dressed, cooked breakfast, walked down to the fort and raised the flag. You started a daily report then made a tour of inspection. There was not a single human voice to greet you while making your daily rounds. A few birds took flight when they heard your footsteps echoing on the granite floor. Small animals and rodents scurried at your approach. Your lantern cast weird shadows on the walls. There was no canteen available so you could drop in for a hot cup of coffee, a doughnut or tobacco for your pipe. There was no P.X. [Post Exchange] *enabling you to purchase writing paper or personal items. You ended your inspection tour...then returned to your little house and prepared an early lunch.*

Cayford continues:

When his daily military chores at the fort had been completed, Sergeant Hegyi would walk down the steep embankment to the river, the site of the Prospect Ferry Store. Here at last, he had the opportunity to hear another human voice. Hegyi allowed himself one glass of beer per day. He would sip his beer, talk with his friends and when he was ready to go back up to the fort, he would say, "Now I've got to go the hill up."

Hegyi billeted in a small wooden house on a corner of the parade ground. He soon turned the officers' quarters inside the fort into chicken coops. Hegyi would collect the eggs and row across the river to Bucksport, where he sold them to supplement his meager salary. The routine of Hegyi's life was occasionally disturbed when teenagers from the town rowed across the river to "capture" the fort and tie up the old sergeant, now well into his sixties.

Sergeant Hegyi was always well dressed in a spotless uniform and tall, black riding boots. (It is said the old equestrian kept a horse in a Bucksport stable, which he exercised regularly.) Hegyi appears to have been especially interested in history and had a personal library of almost two hundred volumes. An inventory of his books included twenty-five volumes of the *Encyclopedia Britannica*, thirty-two volumes of *Abbott's Histories*, eight volumes of the *Library of Universal History*, nine volumes of *A History of the World* and thirteen volumes of *The Great Commanders*.

The Hungarian sergeant was serving at Fort Knox when the Spanish-American War broke out in 1898. The United States' shortest war lasted less than a year. USS *Maine* was sunk in February 1898, war was declared in April and a peace treaty was signed in December.

Fort Knox's involvement was brief; it served as a training center for recruits from Connecticut. Approximately five hundred men from Hartford drilled on the grounds of the fort during June and July. John Cayford summarizes the fort's limited participation: "By the end of the fall [1898], the remaining troops had departed, the mines had been removed from the river and the Navy sailed away. Fort Knox was quiet again."

Even before the war broke out, it was clear Sergeant Hegyi was in declining health. A letter written in 1897 by Colonel R.P. Hughes on an inspection tour of the fort concludes:

Sergeant Hegyi is still in charge. He is looking quite feeble and informs me that he is sixty-eight years of age. It is my conviction that, when the law will

Soldiers of the First Connecticut Infantry Regiment camped at Fort Knox during the Spanish-American War in 1898. *Collections of Maine Bureau of Parks and Land, Courtesy of Maine Memory Network.*

permit this old soldier to lay down his responsibilities by being transferred to the retired list, he should be so transferred. It would be better for the sergeant to spend the remainder of his days with his family in Brooklyn.

Clearly, the army did not follow Colonel Hughes's recommendation, and the next thing we hear of is that Hegyi "was reenlisted" for another three years on October 4, 1897. "A hitch he would not complete," Cayford remarks.

Cayford goes on to describe what happened next: "On July 16, 1900 one of the local fishermen noticed that the flag was not flying from the fort. He investigated and found Sergeant Hegyi extremely ill. He sent word for the doctor who proceeded to the fort on the next ferry only to find the old Sarge unconscious."

When Louise Hegyi was informed her husband was gravely ill, she immediately took the train from New York. By the time she arrived the next day, however, her husband had died.

Ordnance Sergeant Leopold Hegyi, age seventy-one, was buried in the Narrows Cemetery, Sandy Point, Maine.

The Last Hundred Years

Fort Knox continued to be manned by army ordnance sergeants until 1923, when Congress instructed the secretary of war to dispose of a number of excess military posts, one of which was Fort Knox. That same year, Maine's governor, Percival Baxter, was authorized by the state legislature to purchase the property. The agreed-on price was $2,121.

In 1940, the property came under the jurisdiction of the department of parks and recreation, but by 1990, the fort was in such poor repair that the state was unable to pay for the upkeep. The next year, the Friends of Fort Knox was formed, and it has worked with the state ever since to maintain, manage and promote the fort's affairs. Fort Knox is a Maine State Historic Site, and in 1970, it was declared a U.S. National Historic Landmark.

Today, Fort Knox is the most visited historic site in Maine. Decaying roofs have been repaired and exterior lighting installed. The officers' and enlisted men's quarters have been restored, and there is ongoing repair of the masonry. Another Friends of Fort Knox project has been the transformation of the torpedo storage shed into an impressive visitor and education center.

Although there are occasional rumors of ghost sightings on the grounds, the Friends of Fort Knox has yet to take a position on whether the fort is haunted.

OTHER MAINE FORTS OF NOTE

FORT BALDWIN (1905–1912)

Phippsburg, Maine

FORT BALDWIN stands on a hill above the mouth of the Kennebec River. It was built prior to World War I to replace nearby Fort Popham, which had become obsolete. Fort Baldwin consisted of three massive concrete batteries in addition to a five-story observation and fire control tower. The fort was active during World War I and World War II. Fort Baldwin is a State Historic Site and is open to the public.

Fort Baldwin is located on a bluff above Fort Popham. The fort had a garrison of two hundred soldiers during World War I. *Author's collection.*

FORT FAIRFIELD (1839–1843)

Fort Fairfield, Maine

THIS BLOCKHOUSE was named in honor of Maine governor John Fairfield and was built during the Aroostook War in 1839. In 1976, a replica of the wooden blockhouse was built by the Frontier Heritage Historical Society as a bicentennial project. It contains the Blockhouse Museum, which is filled with historical artifacts from early Fort Fairfield, including agricultural items, antiques, photographs and documents. The fort is open to the public by appointment.

FORT FOSTER (1899–WORLD WAR II)

Gerrish Island, Kittery, Maine

FORT FOSTER was one of a number of coastal fortifications built following the Spanish-American War to protect East Coast harbors. In this case, it was built to protect Portsmouth Harbor and the naval shipyard. Today, the fort consists of decaying World War II concrete gun emplacements and bunkers that can be explored. The site was named to honor Civil War general John Foster of New Hampshire. Today, Fort Foster is an eighty-eight-acre town park with recreational facilities and paths running along a spectacular rocky shore. The park is open from 10:00 a.m. to 8:00 p.m. from Memorial Day to Labor Day.

FORT HALIFAX (1754–1766)

Winslow, Maine

FORT HALIFAX was built at the beginning of the French and Indian War, when colonial authorities in Massachusetts realized Maine's vulnerability to attack from Quebec. In 1754, General John Winslow and six hundred soldiers were ordered to build a palisade fort at the confluence of the Kennebec and Sebasticook Rivers. A contemporary drawing of the fort shows two blockhouses, a barracks, a main building and two additional blockhouses on a nearby hill. When a flood washed one of the blockhouses away in 1987, it was carefully restored using many of the original logs that were discovered downstream. Fort Halifax is a National Historic Landmark and is the centerpiece of scenic Fort Halifax Park in Winslow, Maine.

FORT LEVETT (1898–WORLD WAR II)

Cushing Island, Portland, Maine

FORT LEVETT is one of four forts in Casco Bay that were built following the Spanish-American War as part of a coastal defense program instituted by the War Department. By the late nineteenth century, the three older Portland Harbor forts, Preble, Scammel and Gorges, were considered obsolete. In addition to Fort Levett, the list of "new" forts includes Fort McKinley, Fort Lyon and Fort Williams. Each fort guarded the city of Portland from a different vantage point. Fort Levett, on Cushing Island, guarded the main channel to Portland Harbor. The army fortified Fort Levett during both world wars, during which it also served as a telegraph station. Following World War II, Fort Levett was decommissioned. Today, the fort covers 140 acres on Cushing Island, which is occupied seasonally by forty-five families. Access to the island is private.

Fort Lyon (1896–1946)

Cow Island, Portland, Maine

Fort Lyon is located on twenty-six-acre Cow Island in Portland Harbor and is one of the smaller fortifications in the Casco Bay area. Its primary function was to serve as an auxiliary base for nearby Fort McKinley on Great Diamond Island. Fort Lyon's defenses were upgraded during the First World War, when it was occupied by 85 troops. At the start of World War II, an antiaircraft battery was added, and the fort's facilities were expanded to house 150 troops. Fort Lyon was deactivated in 1946. Rippleffect, an adventure education program serving youths in the Greater Portland area, currently owns Cow Island. There is public access along the western shore of the island.

Fort McKinley (1891–1945)

Great Diamond Island, Portland, Maine

Fort McKinley is another of the four forts built on Casco Bay islands to guard Portland Harbor. Originally an artists' retreat, the fort was built on Great Diamond Island shortly before the Spanish-American War. Fort McKinley was an Endicott Period fort, meaning it combined a variety of coastal defenses including breech-loading guns, huge mortars and minefields. Fort McKinley remained in active service until 1945, when it was declared obsolete. Beginning in the 1990s, many of the fort's impressive barracks buildings were restored as private residences. Abandoned artillery batteries still stand on privately owned land in the woods on the eastern end of the island.

FORT O'BRIEN
(1775–1781, 1808–1818, 1863–1865)

Machiasport, Maine

FORT O'BRIEN was built in 1775 following the first naval battle of the American Revolution, which took place off its shores. The fort was initially a four-gun battery that guarded the mouth of the Machias River. The historic site is one of the few Maine forts that were active during three wars: the American Revolution, the War of 1812 and the Civil War. Throughout this period (1775–1865), the fort played an important role in protecting the Machias River and nearby towns. Today, the site of Fort O'Brien is open to the public. In 1923, the United States government deeded the site to the State of Maine. It was made a State Historic Site in 1966 and is maintained by the Bureau of Parks and Lands.

FORT POWNAL (1759–1775)

Stockton Springs, Maine

FORT POWNAL was built by Massachusetts Bay governor Thomas Pownal, who led an expedition to defend the mouth of the Penobscot River in 1759. The star-shaped design was considered exceptional for its time. The fort was never attacked during the French and Indian War, and its presence encouraged English colonists to settle in the area. At the start of the American Revolution, the fort was captured by the British, who seized the fort's guns and ammunition. Shortly afterward, a colonial regiment burned the large blockhouse to avoid its falling into British hands. The excavated remains can be seen today as part of Fort Point Park, which is open to the public. Fort Point Light is located on the edge of the site.

FORT ST. GEORGE (1607–1608)

Phippsburg, Maine

FORT ST. GEORGE was built in 1607 by the Popham Colony settlers near the mouth of the Kennebec River, fifteen miles south of present-day Bath. The fort was abandoned the next year following the death of George Popham. One of the settlers, John Hunt, drew a map of the fort, which was discovered centuries later in a Spanish archive. In 1994, Dr. Jeffrey Brain of the Peabody Essex Museum discovered the site of Fort St. George and began excavations using the Hunt map as a guide. His findings have provided us with a time capsule of life in the tiny early seventeenth-century settlement. Fort St. George is located on a cove between Fort Popham and Fort Baldwin and is an ongoing archaeological site.

Dr. Jeffrey Brain (standing) with volunteers working on the dig at Fort St. George during the summer of 2011. *Author's collection.*

FORT SULLIVAN (1808–1873)

Eastport, Maine

FORT SULLIVAN consisted of a four-gun, circular earthwork containing a wooden blockhouse and barracks. It was built in Eastport, Maine, opposite New Brunswick, Canada. The fort's purpose was to serve as a coastal defense for the easternmost United States during the War of 1812. The garrison surrendered in July 1814, when a fleet of British warships sailed into the harbor. The British occupied the fort until June 1818, when they gave it back to the United States. Fort Sullivan was used as a military facility until 1873. The ruins of the original powder magazine exist today off McKinley Street in Eastport. The former officers' quarters, now the Barracks Museum, was moved to its present location on Washington Street, also in Eastport.

FORT WILLIAMS (1899–1963)

Cape Elizabeth, Maine

LOCATED ON the southern edge of Casco Bay, Fort Williams's origins date to 1872, when it was constructed as an auxiliary base for Fort Preble. The fort was officially named in 1899 by President William McKinley and was armed with twelve-inch Rodman guns. During World War I, it was occupied by artillery companies and National Guard troops—antiaircraft batteries were added during World War II. Following the Second World War, Fort Williams became an administrative center for state military personnel until it was closed in 1963. Today, Fort Williams Park is owned by the Town of Cape Elizabeth and is a popular center for tourism and recreation. The park is open to the public throughout the year.

At the end of his book Maine Forts, *published in 1924, Henry Dunnack provides us with a list of the more than one hundred forts built in Maine from 1607 to 1908. Most are no longer standing.*

BIBLIOGRAPHY

Bradley, Robert L. *The Forts of Maine, 1607–1945*. Augusta: Maine Historic Preservation Commission, 1981.

Bradley, Robert L., Helen Camp and Neill De Paoli. *The Forts of Pemaquid*. Augusta: Maine Historic Preservation Commission, 1994.

Cayford, John E. *Fort Knox—Fortress in Maine*. Bangor, ME: Cay-Bel Publishing Co., 1983.

Chesson, Frederick. "The Forts of Maine." *Sunday Republican*, supplement, August 23, 1987.

Cobb, Ella Waite. *A Memoir of Sylvanus Cobb*. Boston: C.J. Peters & Son, 1891.

Dow, George F. *Fort Western on the Kennebec*. Augusta, ME: Gannett Publishing Company, 1922.

Duncan, Roger F. *Coastal Maine: A Maritime History*. Woodstock, VT: Countryman Press, 2002.

Dunnack, Henry. *Maine Forts*. Augusta, ME: Press of Charles E. Nash & Son, 1924.

Eastman, Joel W. *Harbor Forts: A Look Behind the Walls, 1775–1945*. Portland, ME: Portland Harbor Museum, 2006.

Eaton, Cyrus. *History of Thomaston, Rockland, and South Thomaston, Maine, from Their First Exploration, A.D. 1605*. Charleston, SC: Nabu Press, 2010.

Goodheart, Adam. *1861: The Civil War Awakening*. New York: Alfred A. Knopf, 2011.

Grindle, Roger. *Milestones from the Saint John Valley*. Fort Kent, ME: Project Brave Bulletin, 1977.

Humiston, Fred. "The Aroostook War." *Maine Sunday Telegram*, 1965.

Hunt, H. Draper. *Hannibal Hamlin of Maine*. Syracuse, NY: Syracuse University Press, 1969.

Kaufmann, J.E., and H.W. Kaufmann. *Fortress America*. Cambridge, MA: Da Capo Press, 2004.

Lewis, Emanuel L. *Seacoast Fortifications of the United States*. Annapolis, MD: Naval Institute Press, 1979.

Linz, William. *Maine Forts*. Mount Desert, ME: Windswept House Publishers, 1997.

Longley, Diane G., ed. *Fort Popham, Maine, A Civil War Fort*. Damariscotta, ME: Victoria Print Shop, 1986.

McDonald, Sheila, ed. *A History of Fort McClary State Historic Site*. Augusta, ME: Wilson's Printers, 1993.

McKinnon, Donna Lee. *Portland Defended: A History of the United States Government Fortifications of Casco Bay, 1794–1945*. Orono: University of Maine, 1987.

Oxton, Beulah S. *Major Dickey: Champion of the Madawaska*. Lewiston, ME: Lewiston Journal Company, 1916.

Preston, Richard Arthur. *Gorges of Plymouth Fort: A Life of Sir Ferdinando Gorges.* Toronto: University of Toronto Press, 1953.

Robinson, Willard B. *American Forts: Architectural Form and Function.* Chicago: University of Illinois Press, 1977.

Scroggins, Mark. *Hannibal: The Life of Abraham Lincoln's First Vice President.* Lanham, MD: University Press of America, 1994.

Smith, Joshua M. *Blockhouse and Battery: A History of Fort Edgecomb.* Edgecomb, ME: Friends of Fort Edgecombe, 2009.

———. "Maine's Embargo Forts." *Maine History* 44, no. 2 (April 2009): 143–54.

Trueworthy, Nance, and David Tyer. *Maine's Casco Bay Islands: A Guide.* Rockport, ME: Down East Books, 2007.

Ulrich, Laurel Thatcher. *A Midwife's Tale: The Life of Martha Ballard.* New York: First Vintage Books, 1990.

Youmans, Harold W. *Fort Preble, Maine and the War of 1812.* Riverview, FL: The 1812 Archive, 2009.

ARCHIVAL MATERIALS CONSULTED AT THE FOLLOWING LOCATIONS:

Castine Historical Society, Castine, Maine

Fogler Library, Special Collections Department, University of Maine, Orono, Maine

Fort House, Colonial Pemaquid State Historic Site, Pemaquid Beach, Maine

Fort Kent Blockhouse Library, Fort Kent, Maine

Friends of Fort Knox Library, Fort Knox, Bucksport, Maine

Maine Historical Society, Portland, Maine

Old Fort Western Library, Augusta, Maine

Penobscot Marine Museum Library, Searsport, Maine

Phippsburg Historical Society, Phippsburg, Maine

Robert Anderson Papers, boxes 6, 17, 18, 19, Manuscript Division, Library of Congress, Washington, D.C.

St. George Historical Society, St. George, Maine

Wiscasset Public Library, Wiscasset, Maine

INDEX

ABOUT THE AUTHOR

Harry Gratwick is a seasonal resident of Vinalhaven Island in Penobscot Bay. A retired teacher, Gratwick had a long career as a secondary school educator. He spent most of these years at Germantown Friends School in Philadelphia, Pennsylvania, where he chaired the history department and coached the baseball team.

Harry is an active member of the Vinalhaven Historical Society and has written extensively on maritime history for two Island Institute publications: the *Working Waterfront* and *Island Journal*.

Gratwick is a graduate of Williams College and has a master's degree from Columbia University. Harry and his wife, Tita, have two sons, a Russian daughter-in-law and two grandsons. Visit him at www.harrygratwick.com.

Courtesy of Christopher Krueger.

ALSO BY HARRY GRATWICK:
Penobscot Bay: People, Ports and Pastimes
Hidden History of Maine
Mainers in the Civil War
Stories from the Maine Coast